Human Factors on the Flight Deck

Preface

The primary task of an airline pilot is to bring his aircraft safely from point A to point B. The need to do this economically and comfortably is of secondary importance.

Yet, the knowledge airline pilots must possess related to optimal accident prevention and the avoidance of "human error" still has room for improvement, despite many years of effort.

Simulator training today focuses predominantly on technical system failure, even though this has been identified as the single decisive factor in only about 8% of all safety-critical incidents. It occurs in combination with other factors in another 25% of all incidents, while two-thirds of these incidents are solely related to normal operations, which, at least in recurrent training, is afforded only very little attention.

Pilots should know more about when and under which conditions accidents occur and how they can be sensibly avoided in a preventative manner. Therefore, it is precisely at this point that we focus our attention, providing extensive related examples and explanations in the chapters that follow.

An essential component of accident prevention is safety-relevant behaviour, particularly as conveyed through Crew Resource Management (CRM). Over the past years, this has found broad application in the acquisition of the theoretical knowledge required for the Airline Transport Pilot License (ATPL). Unfortunately, this theoretical content has not been conveyed in the same depth of detail to those who had already obtained their ATPL prior to this time. At the same time, the practical implementation of this material in the cockpit can be further improved upon, as well. Our desire is to support this by initially defining the CRM content as completely as possible, then by subsequently showing how this material can be transferred from the theoretical training into the cockpit by way of the necessary seminars.

Of particular note is that this book was written almost invariably by airline pilots in their technical language and taking into account their daily professional activities. The material is therefore not overly theoretical, but rather straightforward and quite simply "taken from practical application for practical application" and for implementation into the training program.

For this reason, the book should not only be seen as an aid to flight operations in the continuous development of their training programs; it supports the laborious task of determining the content material for CRM training. But it is also of value to the individual practitioner in his self-study efforts and—as a third goal—it enables the German Airline Pilots' Association, "Vereinigung Cockpit" (VC), to formulate its position with respect to the legislative authorities.

Notes

The authors speak for the Vereinigung Cockpit e.V. (VC / German Airline Pilots' Association) and are longstanding members of its Flight Safety working group.

The statements made in this book relate to pilots as a professional title using the *male* gender. It is understood, of course, that they apply to the female gender in equal measure.

This book has been written by German pilots and takes into account neither the interests of an individual flight operation nor those of another cultural framework outside that of Central Europe.

Frankfurt a. M., June 2012

Hans-Joachim Ebermann
Joachim Scheiderer

Contents

Authors

Capt. Johannes Bühler Vereinigung Cockpit e.V., Main Airport Center,
Unterschweinstiege 10, 60549 Frankfurt, Deutschland

Capt. Dr. Gerhard Fahnenbruck Vereinigung Cockpit e.V., Main Airport Center,
Unterschweinstiege 10, 60549 Frankfurt, Deutschland

Capt. Florian Hamm Vereinigung Cockpit e.V., Main Airport Center,
Unterschweinstiege 10, 60549 Frankfurt, Deutschland

Capt. Patrick Jordan Vereinigung Cockpit e.V., Main Airport Center,
Unterschweinstiege 10, 60549 Frankfurt, Deutschland

Capt. Maria-Pascaline Murtha Vereinigung Cockpit e.V., Main Airport Center,
Unterschweinstiege 10, 60549 Frankfurt, Deutschland

Capt. Hans-Ulrich Raulf Vereinigung Cockpit e.V., Main Airport Center,
Unterschweinstiege 10, 60549 Frankfurt, Deutschland

Mrs. Dagmar Reuter-Leahr Vereinigung Cockpit e.V., Main Airport Center,
Unterschweinstiege 10, 60549 Frankfurt, Deutschland

Capt. Rolf Wiedemann Vereinigung Cockpit e.V., Main Airport Center,
Unterschweinstiege 10, 60549 Frankfurt, Deutschland

Other contributors are:

Capt. Tom Becker, Capt. Heiko Blabusch, SFO Jana Feldmann, Capt. Werner Grübel, Capt. Christian Hilgenberg, Capt. Andreas Keller, Reiner W. Kemmler, Capt. Sven Kutschera, Capt. Ralf Nitsche, Capt. Hans Rahmann, Capt. Carsten Reuter, Capt. Carsten Schmidt, Capt. Christopher Selle, Capt. Christina Stromeyer, Capt. Rolf Sulzer.

Abbreviations

AMC	Acceptable means of compliance
ASRS	Aviation safety reporting system
ATC	Air traffic control
BFU	Bundesstelle für Flugunfalluntersuchung (German Federal Bureau of Aircraft Accident Investigation)
CAST	Commercial aviation safety team
CAVOK	Clouds and visibility okay
CBT	Computer-based training
CDM	Central decision maker
CISM	Critical incident stress management
CNS	Central nervous system
CRM	Crew resource management
DLR	Deutsches Zentrum für Luft- und Raumfahrt (German Centre for Aviation and Astronautics)
DME	Distance measuring equipment
EASA	European aviation safety agency
EEG	Electroencephalography
EFIS	Electronic flight instruments system
ENS	Enteric nervous system
FAA	Federal Aviation Administration
FAR	Federal Aviation Regulations
FCL	Flight crew licensing
FCU	Flight control unit
FDM	Flight data monitoring
FMS	Flight management system
FO	First officer
FSF	Flight safety foundation
ft	Feet
GPS	Global positioning system
GPWS	Ground proximity warning system
IATA	International Air Transport Association
ICAO	International Civil Aviation Organization
IFR	Instrument flight rules
ILS	Instrument landing system

JAR	Joint Aviation Regulations
LOFT	Line orientated flight training
MCP	Mode control panel
MEL	Minimum equipment list
MTOW	Maximum take-off weight
NOTECHS	Non-technical skills
NREM	Non-rapid eye movement
NDM	Naturalistic decision making
NTSB	National Transportation Safety Board
OPC	Operator proficiency check
PAPI	Precision approach path indicator
PM	Pilot monitoring
PNF	Pilot not flying
PSR	Perceived safety risk
REM	Rapid eye movement
R/T	Radio telephony
SFE	Synthetic flight examiner
SID	Standard instrument departure
SOP	Standard operating procedure
STAR	Standard instrument arrival route
TCAS	Traffic collision avoidance system
TEL	Translation and elaboration of legislation
TGL	Temporary guidance leaflet
T/O	Take-off
TRE	Type rating examiner
TRI	Type rating instructor
TSA	Time since awake
UAL	United airlines
VASIS	Visual approach slope indicator system
VC	Vereinigung Cockpit (German Airline Pilots' Association)
VFR	Visual flight rules

Accident Prevention

1

Hans-Joachim Ebermann and Patrick Jordan

1.1 Introduction

The following material is fundamental to flight safety. It is a prerequisite to the acceptance of the airplane as a means of public transportation, as well as to the economic viability of a flight operation.

Legislators and regulators are cognizant of the public's interest in safety and establish minimum standards, which must be complied with accordingly.

But minimum standards alone can't sustain long-term flight operations; especially those employing a large number of aircraft. These operators are forced to strive for higher standards then those prescribed by law due to the greater probability of accident or incident. An operator with too many mishaps will be eliminated from the competition. Yet, no airline will invest more money into safety than makes economical sense.

The goal of every pilot is to achieve an accident rate of zero. In the routine of the daily flying profession, the pilot is the final authority for guaranteeing flight safety and the prevention of accidents. It is therefore essential that pilots possess comprehensive knowledge about both the cause, as well as the prevention of accidents. This enables them to function from the outset in a preventative manner.

For this reason, it would be desirable for this material to be included for study and testing, even during initial training for license issuance.

To begin with, it will be beneficial to review some basic flight safety-related statistics. Then the question as to the circumstances under which accidents occur will be approached. This, in turn, will allow us to derive recommendations useful

H.-J. Ebermann (✉) · P. Jordan
Main Airport Center, Vereinigung Cockpit e. V, Unterschweinstiege 10, 60549 Frankfurt, Germany
e-mail: vc.ebermann@onlinehome.de

P. Jordan
e-mail: patrick.jordan@gmx.de

H.-J. Ebermann and J. Scheiderer (eds.), *Human Factors on the Flight Deck*, DOI: 10.1007/978-3-642-31733-0_1, © Springer-Verlag Berlin Heidelberg 2013

for daily air service. Several fundamental psychological principles that work to impede pilots from consistently implementing these recommendations will then be addressed. This, then, leads to the question of "human error", with the subsequent chapters being dedicated to its prevention.

In the context of this chapter, the levels of accident prevention falling under the responsibility of the various authorities will not be addressed or, if so, then only where deemed necessary.

1.2 Accident Statistics

1.2.1 Trends in Accident Rates

In 2007, there were 20,700 jet powered transport aircraft in operation around the world conducting 20.8 million flights. This corresponds to an average of about 1,000 flights per year per jet (see Fig. 1.1) (Boeing 2011).

Accident rates in the USA and Canada are approximately the same as those in the part of Europe regulated by the European Aviation Safety Agency (EASA).

After the heavy losses experienced by the civil jet aviation fleet in the early 1960s, accident rates began to stabilize around the end of the 1990s. From about the year 2000 on, the rate of fatal accidents in North America has virtually dropped to zero.

Several developments at this time correlate with one another. The technical reliability and equipage of Aircraft has undergone continuous development. The operating environment, such as weather forecasting, ATC and the airport infrastructure, has matured. The process of pilot selection and training has been perfected to a greater degree.

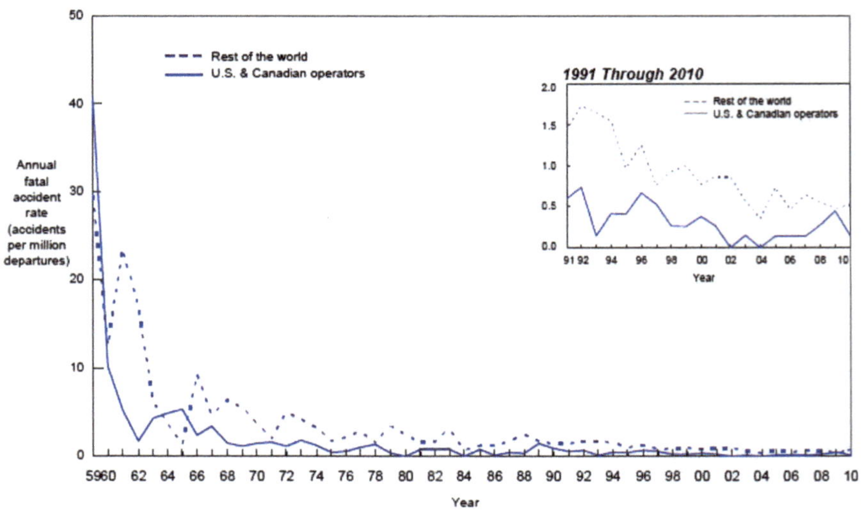

Fig. 1.1 Fatal accident rates (Boeing 2011)

Table 1.1 Accident rates based on aircraft generation (put into service—end of 2008)

Aircraft generation	1st	2nd	3rd	4th	All
Hull loss accidents per million flight cycles	7.46	1.65	0.54	0.34	1.62
Fatal accidents per million flight cycles	4.01	0.86	0.35	0.16	0.87

Table 1.2 Accident rates based on aircraft generation (1998 to end of 2008)

Aircraft generation	1st	2nd	3rd	4th	All
Hull loss accidents per million flight cycles	23.40	3.05	0.57	0.26	1.06
Fatal accidents per million flight cycles	4.87	0.97	0.32	0.08	0.41

Improvements in flight safety related to developments in aircraft technology are illustrated by statistics provided by (Airbus 2009) (see Table 1.1):

The transition from 2nd generation aircraft (e.g. DC-10, Tri-Star) to 3rd generation aircraft (glass cockpit, FMS: e.g. B757/767, A300/310) reduced the number of total losses per 1 million flights by a factor of 3. The transition from 3rd to 4th generation aircraft (fly-by-wire with flight envelope protection: e.g. A318-321, A330/A340, B777) further reduced that number by a factor of 2 (see Table 1.2).

1.2.2 Accident Rates Based on Aircraft Type

This can also be seen in the accident rates per individual aircraft type (see Fig. 1.2).

It can clearly be seen that the aircraft types commonly in use around the world today have only very low accident rates. Among today's modern airlines, however, the MD-11 stands out with an above average rate.

Between 1959 and 2007, a total of 854 aircraft were accounted as *total losses* (hull loss) while 565 *fatal accidents* were recorded with a combined 28,621 deaths. Incidentally, a total of 1,564 *accidents* were recorded during the timeframe referenced in these statistics from Boeing.[1]

---

[1] Unless otherwise noted, these statistics relate to jet-powered aircraft used in civil aviation with over 60,000 lbs. MTOW, excluding those aircraft produced in the former Eastern Block countries, for which reliable data is not available. The following definitions correspond to those from the ICAO. Somewhat abridged, the following defines three categories of accidents:

- Accident: an occurrence resulting in substantial damage to an aircraft or serious injury to a person while persons are on board the aircraft for the purpose of transportation. These statistics exclude acts of sabotage, military attack or attempted suicide, as well as accidents involving freight beyond the reach of passengers and crew.
- Fatal accident: an accident resulting in at least one fatal injury.
- Total loss (hull loss): an accident resulting in repair costs that exceed the current value of the damaged aircraft.

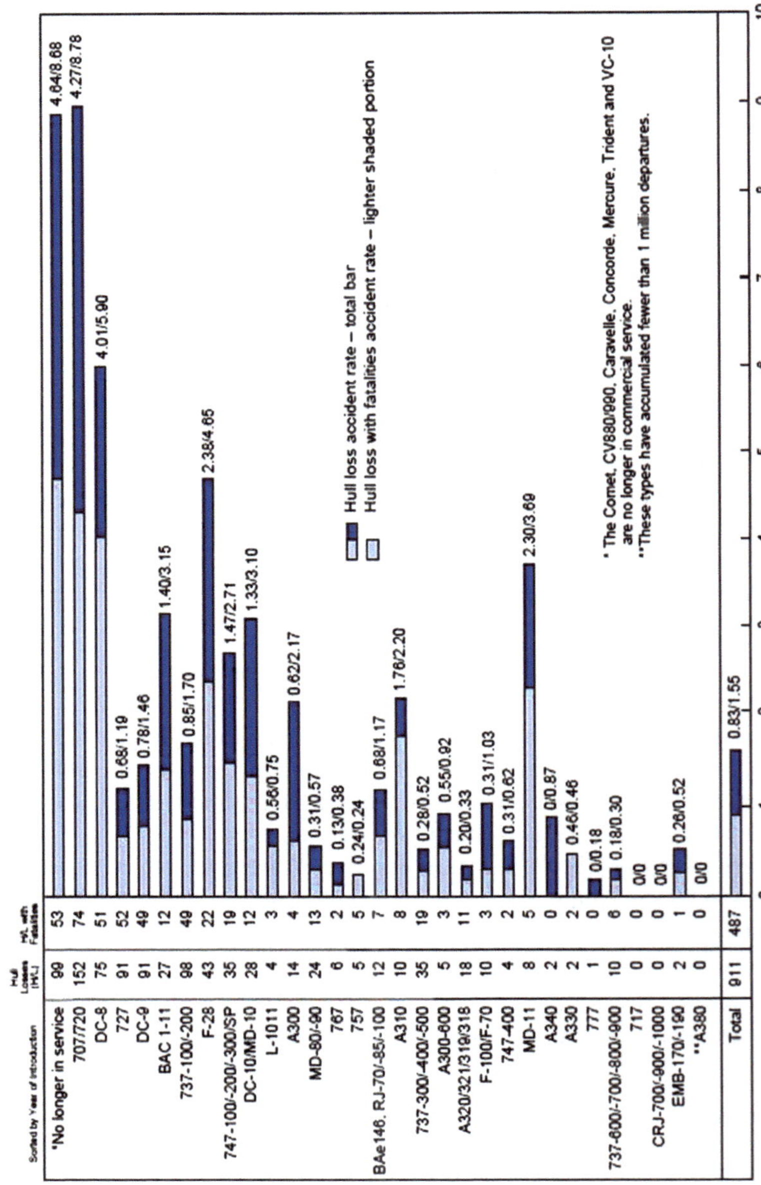

Fig. 1.2 Accident rates based on individual aircraft type (Boeing 2011)

The statistics above, although appearing at first glance to be on the positive side, are put into perspective when one considers that traffic volumes are continuously increasing. This means that, even if accident rates remain consistently low, the ultimate number of accidents will continue to rise.

1.2.3 Distribution According to Traffic Region

This data from the International Air Transport Association (IATA) depicts total loss rates for the year 2008 (IATA 2009) (see Fig. 1.3).

Accident rates in the world's lesser economically developed regions are significantly higher. The reasons for this are manifold: older, poorly equipped aircraft fleets, infrastructure deficiencies and, ultimately, personnel selection and training at oftentimes less than adequate levels.

1.2.4 Accident Distribution According to Type of Operation

It is conspicuous that, in the recent past, considerably more accidents have occurred during charter, freight, ferry, test and maintenance flights than during commercial passenger flight operations (see Fig. 1.4).

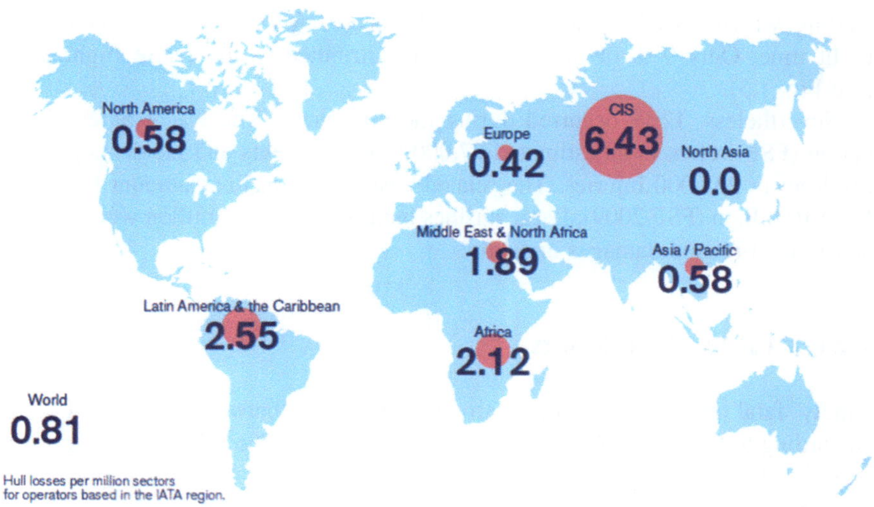

Fig. 1.3 Total loss rates for "western-built jets" according to traffic region (IATA 2009)

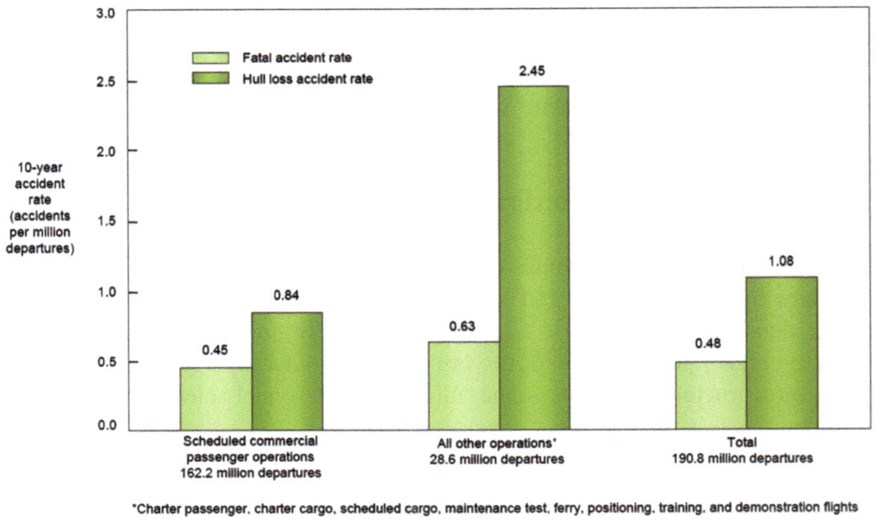

Fig. 1.4 Accident rates according to type of operation (Boeing 2009)

1.2.5 Accidents According to Phase of Flight Phase

The accident-rich phases of flight are takeoff and climb, with 31 % of all accidents occurring within 16 % of the average flight time and, even more so, approach and landing, with 43 % of the fatal accidents also occurring within 16 % of the average flight time. Only 9 % of the accidents are attributed to the cruise flight phase (see Fig. 1.5).

Nevertheless, 12 % occurred during the taxi phase. The Flight Safety Foundation (FSF) refers to an estimated 27,000 ramp accidents and incidents per year resulting in 243,000 injuries. The related costs to the airlines amount to at least USD 10 billion (FSF 2009). Large airlines incur EUR 20–30 million worth of taxi- and ground-related damages annually.

1.2.6 Types of Accidents

Ninety fatal accidents occurred around the world between 1998 and 2007. According to CAST/ICAO, these are distributed among the following categories[2]: (see Fig. 1.6 and Table 1.3).

Some remarks regarding the individual accident types:

[2] CAST (Commercial Aviation Safety Team) is a group made up of flight safety experts from North America and Europe with representatives from the authorities, manufacturers, airlines and pilots' associations. A precise description of the accident categories can be found at www.intlaviationstandards.org (status as of 2009).

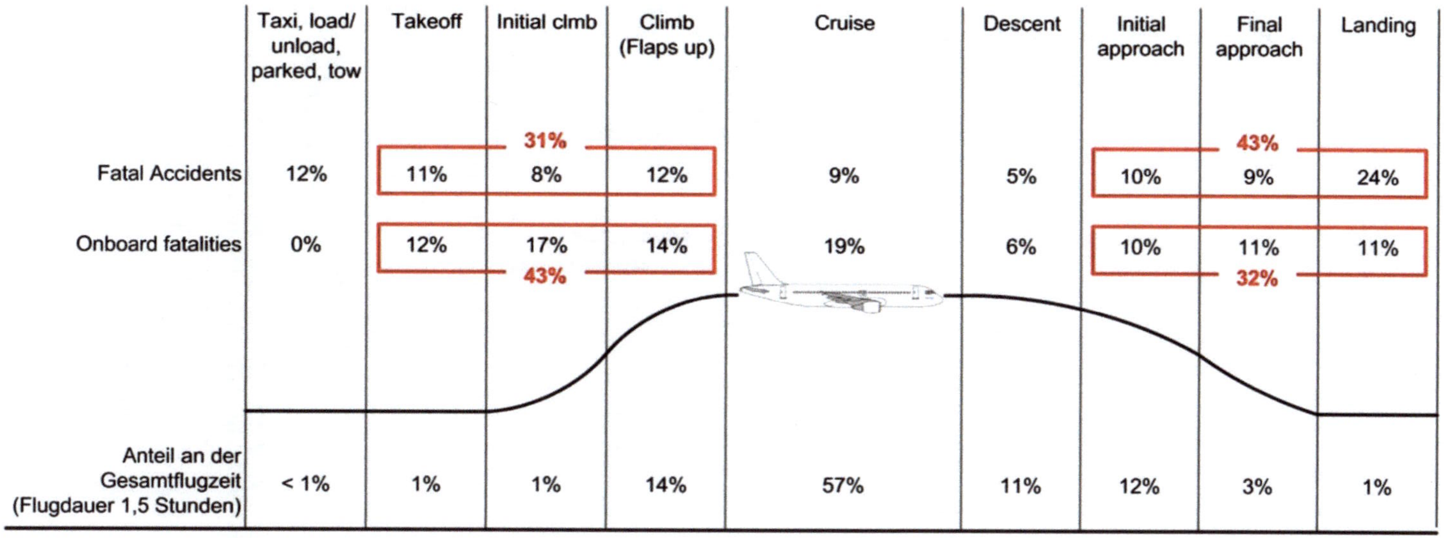

Fig. 1.5 Accidents according to flight phase

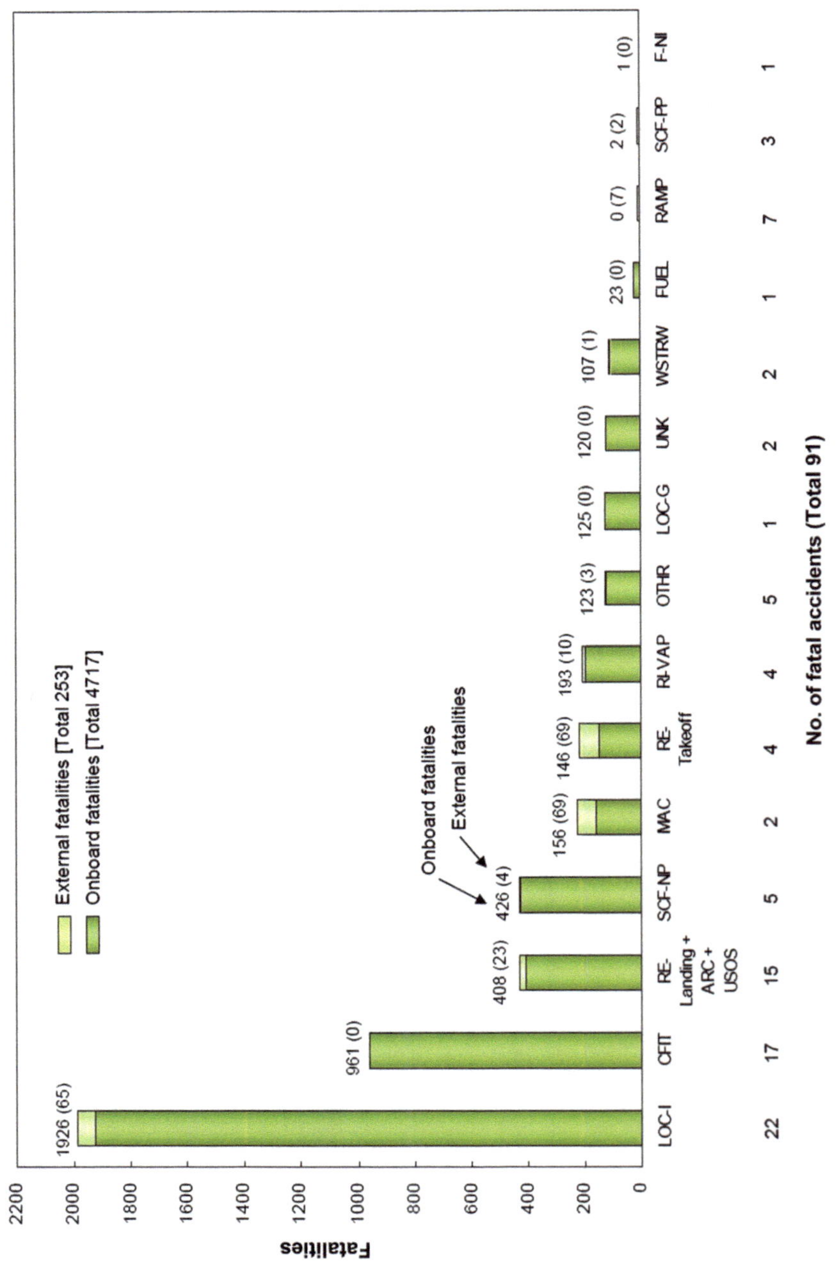

Fig. 1.6 Fatal accident count according to type of accident (Boeing 2009)

Table 1.3 Abbreviations of accident types

Abbreviation	Meaning
LOC-I	Loss of control—in flight
CFIT	Controlled flight into terrain
RE	Runway excursion
RAMP	Ramp handling
SCF-NP	Safety/Component failure or malfunction (non-powerplant)
ARC	Abnormal runway contact
OTHR	Other
RI-VAP	Runway incursion—vehicle, aircraft or person
USOS	Undershoot/Overshoot
MAC	Midair/Near midair collision
UNKL	Unknown or undetermined
WSTRW	Windshear or thunderstorm
SCF-PP	Safety/Component failure or malfunction (powerplant)
LOC-G	Loss of control—ground
FUEL	Fuel related
F-NI	Fire/Smoke—non-impact

CFIT: Almost no aircraft has been lost to date that has had Enhanced Ground Proximity Warning System (EGPWS) installed. It is important to emphasize the ENHANCED function at this point. CFIT accidents, despite being outfitted with "normal" GPWS, headed up the accident statistics for a long time.

Loss of Control in Flight: Aircraft commonly in use today and possessing partial or complete flight envelope protection are only rarely affected by these types of accidents. Yet, they are in no way immune to them if the system is not operated properly (e.g. due to a training deficit) or is defective. Generally speaking, only improved "Upset Recovery Training" can help in this case.

Turbulence: This is the most common cause of injury, albeit not fatalities, in cruise flight.

For modern aircraft, the key accident-related factors continue to be all those associated with the runway: runway excursion, landing, runway incursion.

Table 1.4 Accident rates according to aviation sector

Aviation sector	Flight hours per accident
General aviation	10,000
Business and air taxi flight operations	25,000
Commuter airline flight operations (<30 seats)[a]	150,000
Commercial air transport according to FAR Part 121 (>30 seats) by airlines with a fleet size of up to approx. 100 aircraft	200,000
Airlines with a fleet size greater than 100 aircraft	500,000

[a] until the end of 1997; since 1998, the provisions of FAR Part 121 are applicable from 10 seats.

Table 1.5 Accidents in the German air transport industry (BFU 2008)

Years	95	96	97	98	99	00	01	02	03	04	05	06	07
Accidents	5	8	4	4	6	7	9	6	5	4	6	10	3
Accidents with serious injuries	0	2	1	0	3	3	2	1	2	0	2	5	0

1.2.7 Accident Rates According to Different Aviation Sectors

Table 1.4 depicts the accident rates for the US aviation industry[3] based on flight hours between 1992 and 1997.

In comparison, an automobile accident takes place in Germany almost every 4,000 h. Albeit, it must be noted that the related risk of severe injury or even death per single accident is lower than in an aircraft.[4]

Interestingly: if one considers the relationship of "tons carried per km" with respect to all general means of transport, then the common building elevator proves to be the safest mode of conveyance.

1.2.8 Accidents in the German Air Transport Industry

The German Federal Bureau of Aircraft Accident Investigation (BFU) recorded the following accidents by German registered aircraft (with a maximum takeoff weight > 5.7 t) over foreign and domestic soil (see Table 1.5).

[3] Somewhat more meaningful would be to relate the statistics to takeoffs and landings. Yet, there are no numbers available that apply to the first two columns in the list. The following applies to FAR Part 121 air carriers: average leg length <100 aircraft: 1.2 h; >100 aircraft: 1.7 h, or one accident per 175,000 cycles in companies with <100 aircraft; per 300,000 cycles with >100 aircraft. Numbers ascertained by the Vereinigung Cockpit (German Pilots' Association) stem from NTSB (number of accidents) and FAA (absolute cycles, flight hours) sources.

[4] 2008: 2.3 million accidents recorded by the police with a total of 500 billion (automobile) kilometers driven. Assuming an average speed of 50 km/h means an accident occurs about every 4,000 h. Related injuries and deaths were 400,000 and 5,000 per year respectively (source: German Federal Bureau of Statistics, 2004 Shell study).

The overall accident rate during this period equated to approx. 1 accident per year per 200 registered aircraft. For the professional pilot with a career spanning approx. 30 years, this means that there is an appreciable risk of being involved in an accident during this period.

1.3 Basic Principles

Before presenting the findings from the accident investigations, it is essential to define some of the basic principles of flight safety.

1.3.1 Zero Accident Rate

The reliability of commercial aviation in developed countries has improved to an impressive level since around the year 2000. Yet, the increase in traffic density stands opposed to this positive trend, bringing with it the potential for an increased number of accidents in the future if the rate of accidents per flight is not improved upon even further. In all probability, it will never be possible to achieve a zero accident rate. Notwithstanding, this still remains the elemental goal of every transport pilot.

1.3.2 Safety Net

The share of accidents that could have been prevented by the flight crews is around 70 % (National Civil Aviation Review Commission 1998). When referring to these accidents, one speaks of "human error", and being mostly that of the pilots. In contrast, the number of cases in which the pilots were able to prevent an accident remains uncounted and statistically unrecorded.

Yet, human error, now more commonly referred to as human factors, is very rarely the one and only cause of such accidents. James Reason's model is widely used to explain why this is so. The model assumes a multiple number of preventive levels, where the failure at a single level does not necessarily result in an accident. Only then, when failures take place at multiple levels, can an accident be anticipated (see Fig. 1.7). This is referred to as a safety netor a safety chain (Reason 1991).

Somewhat abridged: Legislators and authorities ensure the creation of and compliance with uniform standards applicable to aircraft manufacturing, crew training and infrastructure. Aircraft manufacturers build the aircraft and its systems. Aircraft operators ensure proper aircraft maintenance, personnel selection, training and compliance with desired operating standards. Pilots function in a preventive manner to ensure as risk-free an operation as possible and, in the event of a malfunction, to defuse the problem in a safe manner. Errors will occur at all of these levels, with the pilots being the last in a long chain of involved parties who could ultimately prevent an accident from happening. They are the final link in the safety chain and are, therefore, often mistakenly seen as the main cause.

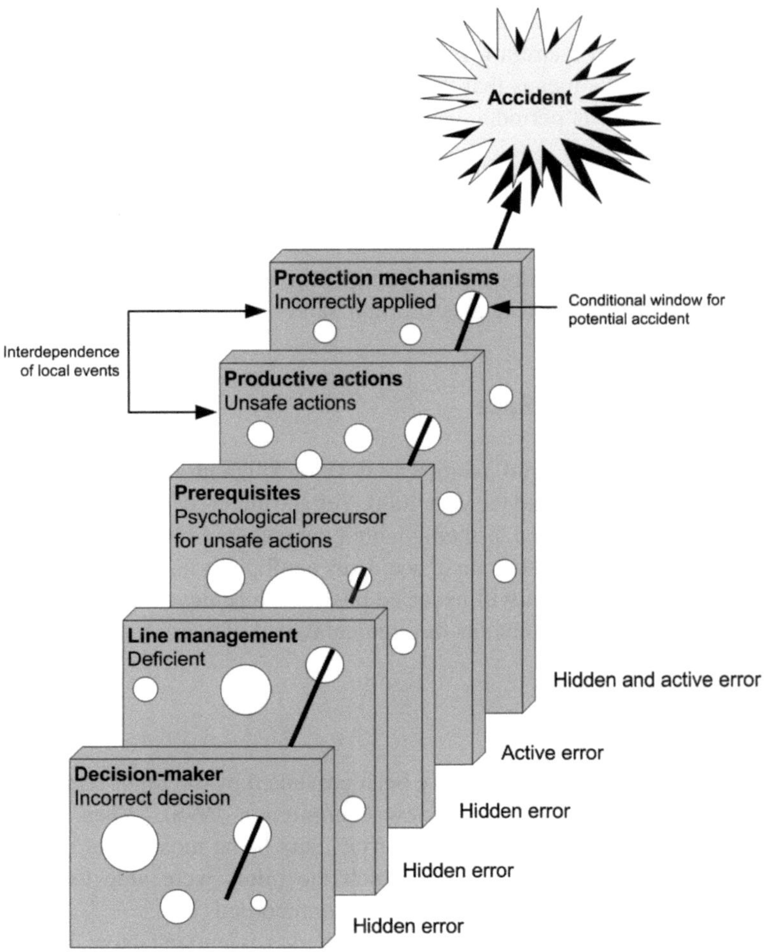

Fig. 1.7 Prevention levels for accident avoidance (according to Reason 1991)

This subject will be discussed in more detail in the chapter on "Human error".

1.3.3 Economics and Flight Safety

If the goal were to achieve a theoretical level of perfect safety, then, in order to limit the effects of an explosion in the cargo bay, for example, the bay would have to be divided into smaller compartments and separated by special and very expensive synthetic, or perhaps even reinforced steel plating. The aircraft would then be "bomb-proof" in the truest sense of the word, but it would weigh 140 tons rather than 40 tons. This, of course, is utopian because it would be extremely uneconomical.

Consequently, flight safety will always be a compromise between hazard, risk and cost.

Statistics reveal that

- airlines in developed countries are safer than those in lesser developed countries.
- passenger-related flight operations are safer than freight-related flight operations.
- large air carriers are safer than smaller air carriers.

The general aviation sector apparently provides a sufficient degree of safety. This branch of the industry experiences one accident every 10,000 h, yet no radical changes have been made in the technical requirements these aircraft are subject to or in the expertise required of the pilots that fly them. Respectively, a large airline with the same statistical level of safety and 650,000 flights per year would have experienced around 100 accidents, of which a third would have included fatal injuries. They wouldn't have a chance of being successful and would necessarily disappear from the marketplace.

Therefore, the legal demands placed on the commercial aviation industry are at an overall much higher level. But large variances can be found in the rate of accidents in this sector, as well, which can't be explained by technology or differences in the operating environments alone. In general (but of course not always), large air carriers offer a degree of flight safety that is higher than that of smaller air carriers. This also applies to the low cost air carriers, some of whom are operating safer than the network carriers (Flouris 2006). Safety is primarily a question of fleet size and not of pricing policy or the airline's position in the marketplace.

The following example should help explain this:

According to Table 1.2, a small airline has an accident every 200,000 flights. Assume this fictitious company employs a fleet of 5 aircraft and each aircraft flies 1,000 flights per year. As such, the company produces 5,000 flights every year, meaning it would have an accident every 40 years. From an economical standpoint, it would be unwise for this small, fictitious company to invest more into safety than necessary to achieve this accident rate.

In comparison, a large airline with 500 aircraft produces 500,000 flights a year. Assuming the same level of safety as with the small company above, this company would experience two to three accidents each year, of which one to two would result in a total loss and one in fatalities. The following example shows just how threatening such an accident rate can be to the existence of an airline:

- ValuJet almost disappeared from the market in the USA after an accident in Florida and subsequently changed its name to AirTran.
- The Turkish company, Birgenair, met with a similar fate following the total loss of an aircraft in the Dominican Republic with German tourists on board.
- Lauda-Air ran into deep trouble following a total loss in Thailand.
- Crossair went through a severe crisis following a series of three total losses.
- The Cypriot carrier, Helios, also went through a very difficult crisis following a spectacular total loss, which was traced back to deficits in their safety program.

One serious accident can threaten the virtual existence of an air carrier. Consequently, a level of safety must be achieved that can guarantee the carrier will only infrequently be the focus of the public's attention. Looking into the future of the industry, this can also be applied to airline alliances or cooperations who share a common image. Additional investments made into aircraft technology or pilot selection and training is economically wise, because an accident would inevitably be accompanied by a loss of trust and a subsequent rise in revenue shortfalls, neither of which are covered by insurance.

An airline will become increasingly averse to risk as its fleet size increases. They will strive to ensure that fewer and fewer risks are taken in all relevant areas. They will invest more money into safety and assume a short-term economic disadvantage compared to the smaller carrier. This presents a problem in that, while the cost of a particular initiative taken to increase safety can be clearly measured, the anticipated effects of a reduced probability of accident cannot be easily quantified.

1.3.4 High-Profile Accidents

People always seem to be more interested in the large, sensational catastrophes than in the small, everyday accidents. "Small" accidents receive merely marginal attention. Even though German registered aircraft experience 3–10 accidents a year, the public perceives only unique events.

This selective perception of accidents with a large media profile has the effect of forcing air carriers to invest a great deal of time, effort and money into safety. Because of this, passengers on commercial airliners are transported at an objective level of safety that the automobile driver or even a private pilot wouldn't deem necessary.

The safety image of an airline is of tremendous importance: If it is good, one accident—under certain circumstances—could be absorbed without great economic consequence (e.g. the Swissair accident by Halifax). If it is poor, on the other hand, mere speculation could be enough to cause the company serious difficulties. Public opinion does not wait for years in anticipation of the official accident report; it is very quick to issue a premature verdict (e.g. the U.S. airline, TWA, accident by Long Island).

The importance of an airline's safety image is also affirmed by the concept of the public Perceived Safety Risk (PSR) (Simon and Mitchell 2009). This correlation is also referenced in the ICAO Safety Management Manual (see Fig. 1.8).[5]

An airline having achieved the highest PSR "Surplus" level has two options:

- Quality leadership (create a premium market)
- Cost leadership (reduce costs)

By seizing the second option, however, the PSR will drop one level down to "Acceptable". At this level, competition is carried out through ticket price or service. In cases where the standing in public opinion is poor, safety inputs, alone, can lead to a reassessment and ultimately improve the image.

[5] ICAO Safety Management Manual (Doc 9859, 1st Edition): "1.3.3. The air transportation industry's future viability may well be predicated on its ability to sustain the public's perceived safety while travelling. The management of safety is therefore a prerequisite for a sustainable aviation business.".

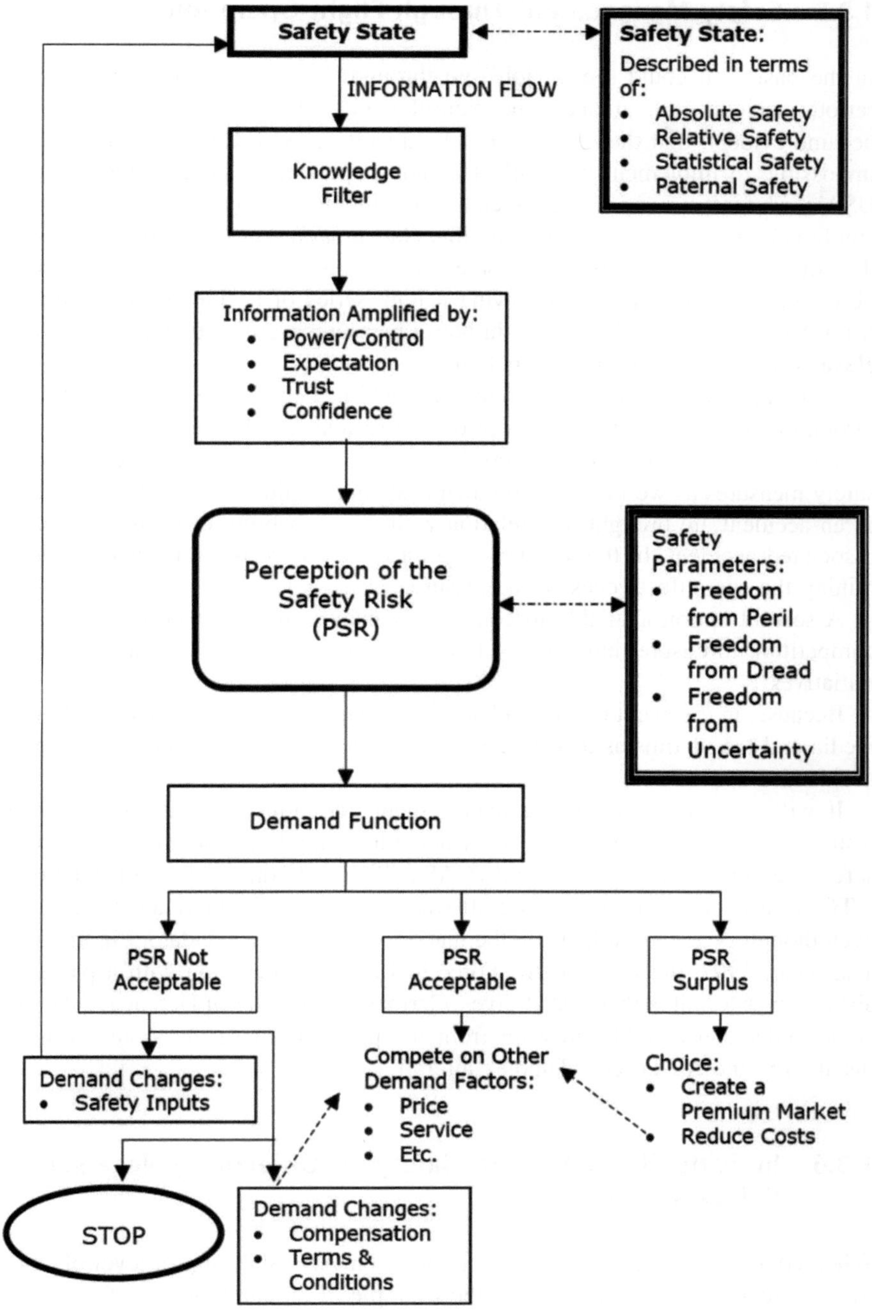

Fig. 1.8 Public perceived safety risk

1.3.5 Safety Management Through Flight Operations

In the past, you could see a "jolt" go through the affected airline following a serious accident. All at once, the painful reality of gaps in the safety culture became evident, that then had to be subsequently closed. Changes that had been impossible to implement were all of a sudden possible. One example of this is USAir, who, after a series of accidents from 1989 to 1994 (five total losses with a combined total of 211 passenger fatalities) began an in-house "revolution". Since that time, it has faded from the safety discussion. A similar situation can be observed with Korean Air. Following a long series of accidents (five accidents from Aug. 97 to Dec. 99), drastic in-house measures were taken that successfully elevated the airline out of the limelight.

From a purely economic standpoint, it would be better to close any safety gaps through more effective prevention prior to a serious accident.

Within the framework of a Safety Management System the costs related to the safety measures as well as the estimated costs that could be expected in the event of an accident are brought into relation with the probability of occurrence of the associated accident. In this manner, a point of reference is established for determining the cost effectiveness of a measure.

A second option available to company management would be to observe the competition to ensure they are not found lacking with respect to safety-relevant initiatives.

Because, in the event of an accident, the management will be questioned by the media and the victims as to whether it did everything in its power to prevent the accident.

It will be problematic if the management consciously saved money by not installing a market-ready system or by not sufficiently implementing a recognized safety measure. One example is the TCAS collision alerting system: Installation of a TCAS system into transport aircraft was not required by German law in 1997, even though systems available on the market were already mandatory in the USA. That year, 1997, the German Air Force had a midair collision with a passenger airliner over Namibia that cost 70 lives. Germany's Minister of Defence at the time came under considerable pressure from the public because the system was not installed in the Air Force's Tupolev aircraft.

1.3.6 Individually "Sufficient" Safety vs. Objectively Necessary Safety, Part I

When controlling an aircraft, the pilot unconsciously establishes a level of safety by "gut feeling" or instinct. Using the example of an automobile, safety is only one aspect out of many the interests the driver while at the controls.

- Driving too fast, and thereby unsafely, can result from deadline pressures or perhaps just the fun of driving fast.

- A small child in the car or a ringing mobile phone can distract the driver's attention.
- Violating minimum traffic separation distances may be accepted as a demonstration of dominance or power.

Inexperience, insufficient routine, negligence, time pressure, dominant behaviour and laziness are always along for the ride and increase the risk, yet they are accepted by society at the cost of avoidable traffic victims.

The integrity of life and limb, as one of the supreme human rights, is permanently disregarded in everyday life through thoughtlessness or carelessness. The fact that transportation ministers do not have the necessary political capacity to take action against this mechanism is demonstrated by the regularity, with which sensible recommendations from experts for improving road traffic safety are buried in the sand. For example, it is still possible to drive a car in Germany with significant concentrations of alcohol in the blood. Yet, alcohol and flight duty are absolutely incompatible for the transport pilot.

It would be an objective necessity to defuse identifiable risks prior to the occurrence of an accident and, in the case of an accident that has already occurred, that it not be allowed to happen again. This is precisely what is demanded of commercial aviation.

All accidents are investigated and evaluated by a national accident investigation body. Recommendations at the end of the accident report should impact all involved aviation stakeholders so that the mistakes identified are corrected.

If this method were applied to road traffic, it would mean the following: an accident due to excess speed, alone, would, as a minimum, result in a recommendation for stricter speed limits. All motorists would adhere to the provision out of conviction. This is obviously an unrealistic scenario. The following section on "Standard Operating Procedures" (SOP), the difference between perceived, individually sufficient and objectively necessary safety will be dealt with in greater detail.

1.4 Origin and Prevention of Accidents

It is possible to derive recommendations towards the prevention of accidents from the knowledge gained about their origin. While every single accident is carefully investigated and evaluated on the one side, there are only a few studies dealing with the commonalities of accidents on the other. Two of these studies are noteworthy: One from the U.S. National Transportation Safety Board (NTSB) and one from Lufthansa.

1.4.1 The NTSB Study

The NTSB analysed 37 commercial aircraft accident reports it had issued during the period between 1978 and 1990. In all the accidents investigated, the pilots were named as the initiating or contributing factor (NTSB 1994). In these 37 accidents, the crews made 302 work-related errors (see Tables 1.6 and 1.7). The number of errors per accident lies between 3 and 19, with an average of 7.

Figure 1.9 depicts the distribution of errors attributed to the respective pilot position. Errors attributed to the flight engineer are excluded.

Table 1.6 Accident-related crew errors

Type of error	Absolute frequency	Relative frequency [%]	Number of accidents per error type
Primary error			
Aircraft handling (AH)	46	15.2	26
Communication (CO)	13	4.3	5
Navigational (NA)	6	2.0	3
Procedural (PR)	73	24.2	29
CRM (RM)	11	3.6	9
Situational awareness (SA)	19	6.3	12
Systems operation (SO)	13	4.3	10
Tactical decision (TD)	51	16.9	25
Secondary error			
Monitoring/Challenging (M/C)	70	23.2	31
Total	**302**	**100**	

Table 1.7 Description of the error types

Error type	Description
Aircraft handling	Failure to maintain the aircraft within defined parameters
Communication	Mistaken readback, misunderstanding, holding back information
Navigation	Incorrect frequency selection, misinterpretation of flight charts
Procedural	Omitted or incorrect callouts, mistaken checklist readouts, non-use of prescribed checklists, omitted or incorrect briefings, omitted information acquisition
CRM	Incorrect workload management, incorrect task priorities, too great a level of dystress, incorrect or omitted transfer of aircraft control
Situational awareness	Aircraft control according to incorrect parameters
Tactical Decision	Failure to make decisions despite clear action signals, disregard of warnings/alerts
Monitoring/Challenging	An incorrect action not observed or not addressed by the other pilot

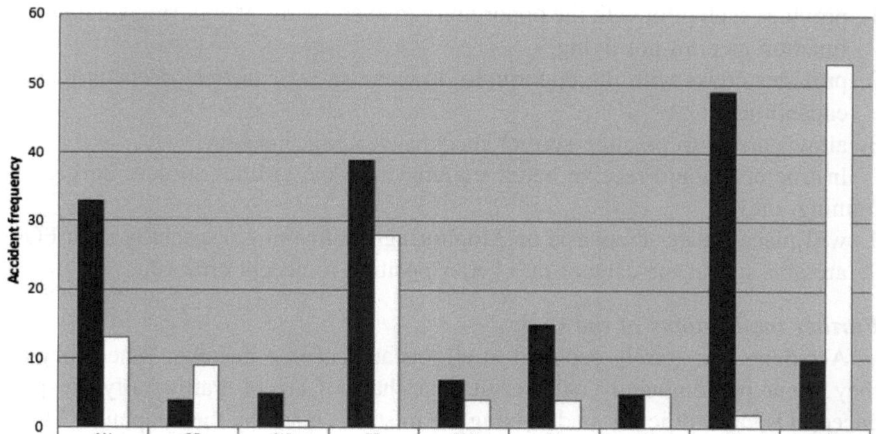

	AH	CO	NA	PR	RM	SA	SO	TD	M/C
■CPT	33	4	5	39	7	15	5	49	10
□FO	13	9	1	28	4	4	5	2	53

Fig. 1.9 Pilot error distribution

Commonalities between the accidents investigated

- In over 80 % of the accidents, the captain was the pilot flying and the first officer (FO) was the pilot not flying.
- The primary failures attributed to the crews were their mistakes in the application of SOPs, incorrect tactical decisions and errors in monitoring (Monitoring/Challenging).
- Errors in Monitoring/Challenging took place in over 80 % of the accidents. *This failure was attributed exclusively to the FO.*
- In 40 % of the accidents, the captain made incorrect decisions that were not challenged by the FO. In most cases, the decisions in question had to do with a failure to follow a required course of action, such as a go-around.
- 55 % of the accidents were on flights affected by a flight delay. The average rate of delay for the overall air transport industry at the time was about 25 %.
- Crews allow themselves to be pressured by delays and make significantly more workload-related mistakes, especially on the ground during flight preparation and taxiing.
- 73 % of all accidents happened on the first day of a joint tour by the captain and FO. A total of 44 % actually happened on the very first leg of the tour.
- Half of the crews were awake for longer than 12 h at the time of the accident (Time Since Awake, TSA). These fatigued crews made significant mistakes in the areas of SOPs and decision making. Especially overnight flights are more frequently prone to accident.
- 53 % of the FOs were in their first year with the company. The average flight time for the FOs on their respective aircraft type was 419 h.

As a consequence, the NTSB required:

A LOFT (Line Orientated Flight Training) component should be scheduled in the simulator for each type rating, which

1. provides each pilot with the opportunity to exercise his Monitoring/Challenging function as pilot not flying,
2. provides crews with the opportunity to exercise their tactical decision-making capabilities,
3. allows crews to practice correct checklist reading procedures.

Instructors should receive better training from the airlines so that, during line training, they

1. will place greater emphasis on Monitoring/Challenging, especially with FOs,
2. are able to put the captain in a better position to accept criticism.

Further implications of the study

Accidents very rarely occur, if at all, because of one mistake. When they do, they occur predominantly as a result of a chain of errors. Particularly inexperienced FOs have difficulties addressing the mistakes made by their captains. This is especially true when they are still getting to know each other and factors such as time pressure and fatigue begin to aggravate the situation. The "novelty of the task" increases the probability of an error by a factor of 17 and "time pressure" by a factor of 11. Captains must know this so they don't overburden their (inexperienced) FOs and, in so doing, deprive themselves of their only source of feedback for recognizing and correcting their own mistakes.

FOs must receive training that puts them in a position to properly assume their role of monitoring the captain, beginning with the first leg alone with a captain during line training. They must be familiar with all safety-relevant SOPs, have the skills to safely handle the aircraft in every phase of flight, be knowledgeable of the aircraft's flight limitations and be prepared to openly address these at all times, even when their thoughts are yet unclear. They must intervene promptly when required and take over the controls as necessary.

- Boeing and several airlines have taken this into account and have replaced the term PNF (Pilot Not Flying) with a more sensible PM (Pilot Monitoring).
- Captains must be able to accept criticism from their FOs and beware of belittling what they deem to be improper or exaggerated criticism.
- Captains should call for criticism anytime he suspects it is being withheld.
- A captain should give his FO an opportunity to become accustomed to him on their first day together, and especially on the first leg of a joint tour. At the same time, risks of any kind should be avoided to the greatest extent possible. This may mean forgoing a voluntarily shortened approach, a visual approach or an "immediate T/O" in order to preclude overburdening his monitoring function.

Even on the first leg, the FO must possess as much self confidence in himself and his capabilities as needed to be able to immediately and openly address disagreements and mistakes.

> The calling for criticism by the captain and the offering of criticism by the FO are crucial to the successful prevention of accidents.

Only then can a hierarchical gradient exist that guarantees the safe working relationship between both pilots. By forcing an FO into a "passenger role"—consciously or unconsciously—the captain potentially deprives himself of an important source of competence, good ideas and problem-solving recommendations.

1.4.2 The Lufthansa Study

To draw any conclusions regarding the state of aviation safety based on relatively few accidents is very difficult from a statistical perspective. In order to improve the statistical basis for determining the impact on operations, it makes sense to not only investigate accidents, but to investigate close calls or safety–critical incidents, as well. For this reason, Lufthansa carried out an intensive study of incidents, which occur much more frequently than accidents, from 1997 to 1999. 2070 pilots took part in the study (Lufthansa 1999).

As it turned out, 99.9 % of the pilots had experienced at least one safety–critical incident in their careers. A surprising quota of approx. 3,000 incidents per year emerged. This equates to 8 incidents per day. Expressed otherwise: Every Lufthansa pilot experienced around one incident per year! In order to narrow the focus and scope of risks, errors were classified according to the following criteria and allocated individually or in combinations of up to four:

- OPS: operational problems
- HUM: human work-related errors
- TEC: technical faults
- SOC: social climate among the crew

Incidents attributed to only one group pose a small risk because a structured cockpit working environment will defuse individual errors. The combination of OPS + HUM + SOC stands out conspicuously; composing 37.8 % of all incidents (see Fig. 1.10). A possible scenario could look like this: An operational problem (OPS) causes an increased workload from which a work-related error (HUM) results that is not corrected due to a stressed cockpit environment (SOC).

It is apparent that a negative social climate acts like a "turbocharger" for accidents. This study showed for the first time that a quantitative correlation can be measured between social climate and flight safety.

The TEC (technical faults) and OPS (errors resulting from operational processes) categories are the least prevalent and can be influenced only to a limited extent by the pilots.

The OPS category deals primarily with bad weather and dangerously close encounters with other aircraft, or so-called "near misses".

Technical faults are comprised mostly of engine and landing gear problems, as well as false indicator readouts on the flight guidance instruments.

Incidents not influenced by pilots in this study make up merely 13 % (1.2 % + 7.7 % + 4.1 %) of all incidents from (TEC + OPS). Conversely, 87 % of all incidents could have potentially been defused by the flight crews.

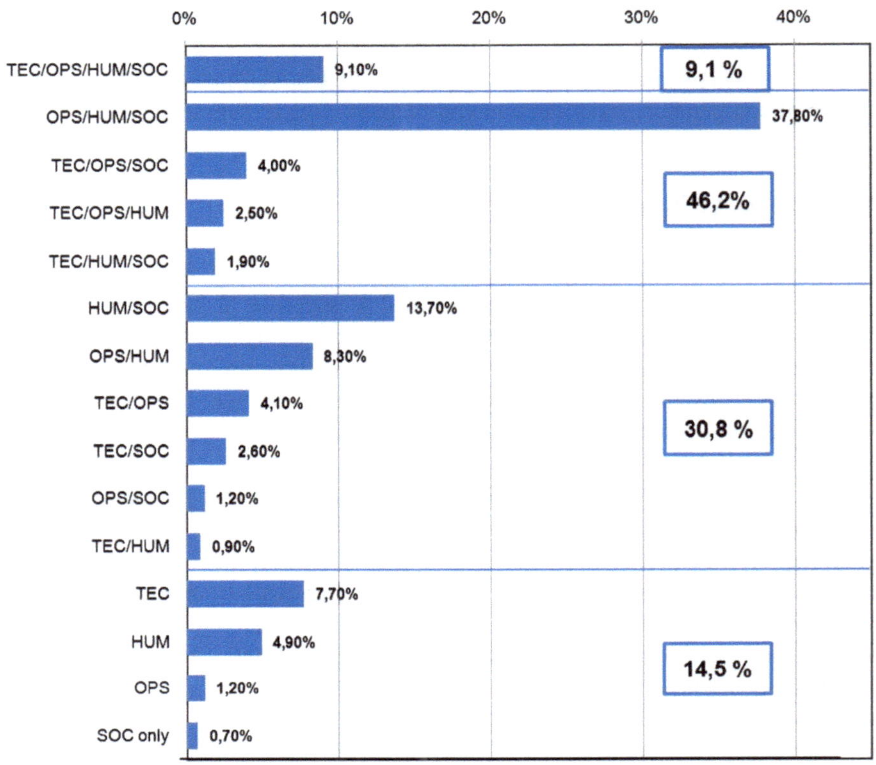

Fig. 1.10 Frequency of event configurations

It is evident from this chart that SOC-related problems played a role in 70 % of all incidents. In all incidents where pilot error was involved, the proportion even rises to 80 %. From the "turbocharger-insight" above, it follows that: 80 % of all incidents, in which human error played a role, could have been prevented if an optimal cockpit environment had prevailed.

The following insight is particularly interesting: As opposed to the common perception that pilots might be overwhelmed by new technologies, this study revealed that the combination of technical problems and human error (TEC + HUM) remained below 1 %, whereby each category by itself resulted in 7.7 % (TEC) and 4.9 % (HUM) respectively.

The analysis revealed the following key aspects:

Communication

Interpersonal communication problems occurred in 53 % of all incidents. Of these, approx. 30 % took place inside the cockpit while 70 % took place between the crew and outside parties. In this context, ATC communications were most often involved, playing a role in 27 % of all incidents. The main points of focus were:

- Mandatory statements, such as callouts in the event of deviations, being omitted.
- Concerns not being expressed.
- Important statements being incomplete, unintelligible, missed or ignored.

Defence strategies to ward against these errors have long been known, yet crews evidently have difficulty applying them with any consequence. When one considers its role in 53 % of all incidents, then a further discussion as to the need for professional communications training should be unnecessary. Here is a short look ahead to the chapter on Communication:

- When something is unclear (cockpit, ATC): Seek clarification from the message sender.
- Use standard R/T consistently.
- Address all deviations without ambiguity.
- Be alert for non-verbal signals.
- First dial in the value, then provide the readback in R/T through the FMA/MCP/COMM.
- Employ the "Sterile Cockpit" concept (80 % of the incidents in 7 % of the time).

A crew member going it alone

A "non-jointly coordinated action" was involved in 12 % of all safety-related incidents. The "lone warrior" syndrome is still a central problem in the cockpit. In most cases, it does not involve "ill will or even a decision made solely by one person". It is more often target fixation under difficult operational conditions that turns a good team player into a solo pilot: a tight slot-time, an expiring hold-over time, the desire to get the passengers to their destination on time.

It lies in the nature of the industry that the problem of a crew member "going it alone" will usually be triggered by the captain. It is easier for the captain to stop an FO from acting alone just based on his hierarchical position and overall responsibility, as well as age and experience. Co-pilots will commonly try to excuse their actions on the incident report by noting: "The captain would have probably acted as he did, regardless".

The study revealed that, in 918 out of a total 1,897 incidents, the FO did not express any criticism. In 210 cases, concerns were expressed, but these were disregarded by the pilot flying. Recommendations stemming from the study are:

- Uneasiness, differing opinions, deviations and objections should be articulated loud and clear.
- Avoid rushing; don't allow yourself to be pushed; create some free space as a buffer for any unforeseen circumstances (fuel, descent, ground times, etc.).

It should be noted at this point:

It is crucial for pilots to maintain a good overview (so-called situational awareness). Thorough flight planning with a deliberate assessment of potential risk helps improve this overview and prevent subsequent problems from arising.

Specific risks should be addressed during the departure, takeoff and approach briefings.

Human work-related errors (HUM)

The effects of work-related errors can be avoided or minimized through a tightly woven safety net and a structured, uniform work routine.

Nevertheless, human error was a factor in 87 % of all incidents. Of these,

- 90 % involved available facts that were not considered,
- 79 % involved the cockpit crew being implicated from the onset, while
- 77 % of the work-related errors were associated with rule violations

It should also be noted at this point that errors are unavoidable; they can't be prevented entirely. Errors are not necessarily safety-relevant as long as they are discovered and caught, such as through a checklist or through intervention and feedback from a colleague. It first becomes critical when errors go undiscovered and evolve into an error chain, which can lead to an incident or accident.

It is important for every transport pilot to analyse work-related errors when detected, either by himself or together with the crew, to avoid repeating them where possible. If it is not possible to address a situation directly when it arises, then a short discussion in the cockpit following landing may be sufficient. It does not have to be long and can be introduced with the questions: "Did you notice any mistakes?" or "Would you have done anything differently than I did?"

Standard Operating Procedures (SOPs)

The most logical starting point for improving flight safety is through disciplined compliance with the SOPs. This is generally well known and has been trained intensively for many years. But then, why is it so hard for the crews to comply with them?

According to the study, the importance of SOPs is generally not called into question by the crews. Nevertheless, they are breached over and over again, either knowingly or unknowingly. With over 2,000 flights a day in a large airline, tight limits are necessary for economic survival, yet these limits must also be padded with clear-cut buffers. The buffers must be available in the event of unforeseen circumstances.

By the same token, mutual monitoring according to a definite set of rules is essential. When limits are transgressed, the gate falls away for the person being monitored—a second limit doesn't exist. Moreover, when a rule violation is tolerated once, the inhibition towards further transgression falls away with it. This encourages entry into the error chain.

> Every SOP that is ignored can represent the last level of prevention prior to the accident.

Unstabilized approach

The unstabilized approach plays a role in 20 % of all incidents and, as such, makes up the lion's share of SOP violations:

- 58 % too high
- 57 % too fast
- 27 % due to lateral offset

- 17 % too low
- 21 % due to incorrect configuration

All these situations could have been elegantly and safely alleviated with a go-around procedure. Yet, the study revealed just how poorly developed the disposition towards this solution really is.

The study also revealed that the majority of unstabilized approaches were carried out by the captain. Instead of the prescribed callouts when a limit is about to be transgressed, it is obvious that the co-pilot's individual tolerance threshold will determine the size of the mesh in the last safety net. This means that the co-pilot is the last resort for containing the error when a captain transgresses the limit.

Deviation from ATC clearances

This error ensues almost always unconsciously, illustrating just how great the various pressures normally are inside the cockpit. This type of error is found in 19 % of all incidents. Of these:

- 45 % is attributed to flying at an uncleared flight level
- 22 % is attributed to course deviation
- 21 % is attributed to deviation from a SID or STAR
- 10 % is attributed to takeoff or landing without clearance

Strategies for error prevention:

- Distractions, communication and unnecessary work must be avoided whenever possible during the critical phases of flight.
- All pilots in the cockpit must hear a clearance.
- Uncertainty regarding ATC clearances should be clarified with the controller, not in the cockpit.
- First enter the value into the FCU/MCP, then readback the value from the display via R/T.

Basic flying

The study revealed that deficiencies existed in flying ability, as well: Problems with basic flying played a role in 25 % of all incidents.

- 60 % occurred during correction of target parameters: too much, too little, too late, too slow
- 33 % occurred during landing: too far, too hard, incorrect flare, deviation from the centerline
- 21 % occurred while taxiing: too fast, using the wrong taxiway and taxiing over runway holding points
- 10 % occurred during go-around: incorrect manoeuvre sequence, dropping below minimum speed and, in three incidents, even contact with the ground

Particularly the go-around, because it appears so infrequently as an incident, is over-represented by a factor of at least 27. This revealed a discrepancy with regard to reliance on the simulator, where it presents no problem. In practice, however, there are large emotional hurdles and even feelings of personal failure to deal with, which may significantly complicate the overall manoeuvre.

An important corrective measure would be to augment "stick-and-rudder training" in the simulator. In so doing, the desired degree of competency can be achieved while a sufficiently rapid "instrument scan" is acquired. The long-haul fleet is especially prone to this problem.

A note from the authors: If it is possible to do so without impairing safety (weather, traffic, ATC, fatigue), increased line operations should also be flown using a reduced degree of automation in an effort to supplement simulator training.

Taxi incidents are all too frequent. The seconds saved by taxiing fast can scarcely be measured; the number of related incidents, however, all the more.

Many FOs do not intervene with callouts, but assume a front-seat passenger mentality "I don't like it either when someone interferes while I'm driving my car". Personal discomfort in this context is a strong indicator that verbal intervention is call for.

Another note from the authors: Where technically possible, it is helpful for FOs to taxi from time-to-time in their role as monitors.

Equipment operation

The crew is well familiar with their airplane and its functions; instrument inputs are repeated daily, a hundred times over. A great deal of self discipline is required to keep from getting complacent in this regard. Only in this manner can operating error, which accounts for 18 % of all incidents, be avoided:

- 30 % is attributed to incorrect inputs
- 20 % is attributed to mistakenly omitted component actuation
- 14 % is attributed to actuating the wrong switch
- 14 % is attributed to actuating the wrong mode

Countermeasures:

- Deliberate verification of inputs into the FMA and compliance with FMA callouts
- The other crew member should also check the result.

1.4.3 The Boeing Study

What can be done to prevent accidents? This question was posed by the American airplane manufacturer, Boeing. They examined 232 accidents that took place around the world from 1982 to 1991 involving transport aircraft with maximum takeoff weights greater than 60,000 pounds. The objective was to determine the prevention strategy in each individual case, which would have averted the accident at its origin. The number of possible strategies ranged from 1 (in 39 accidents) to 20 (in one accident), with an average of barely four strategies per accident (Boeing 1993).

Table 1.8 shows the proportion of the 232 accidents that could have been prevented by the respective strategy. The terms used are understood by flight crews around the world

Table 1.8 Prevention strategies

Strategy	%
Pilot Flying adherence to procedure	42
Other operational procedural considerations/CRM training	38
Embedded piloting skills	25
Pilot Not Flying adherence to procedure	23
Design improvement	21
Maintenance	19
Captain exercise of authority	16
Approach path stability	15
Go-around decision	14
ATC	13
Eliminate runway hazards	11
FO crosscheck—performance as Pilot Not Flying	11
Weather information and accuracy	8
Response to GPWS	7
Airport services	6
Pilot Flying awareness and attention	6
ATC/Crew communications	5
Pilot experience in aircraft type	5
Pilot Flying communication or action	5
Pilot Not Flying communication or action	5
Crew fatigue	4
Fire and rescue services	4
Captain's crosscheck—performance as Pilot Not Flying	3
Training for abnormal condition	3
Availability of approach aids	2
Manufacturing process	2
Performance data	2
Pilot incapacitation	2
Use of all available approach aids	2
Weight and balance control	2

The individual terms are explained in Table 1.9:

The Boeing study largely confirms the findings of both the NTSB and the Lufthansa studies referenced earlier.

Table 1.9 Explanation of the prevention strategies

Strategy	Explanation
Pilot Flying adherence to procedure (42 %) and Pilot Not Flying adherence to procedure (23 %)	The accident could have been prevented through compliance with published procedures.
Other operational procedural considerations/ CRM training (38 %)	Other factors than those identified here, over which flight operations management exercised direct influence. The study explicitly emphasized the importance of improved CRM training.
Embedded piloting skills (25 %)	Improved basic flying and technical skills would have prevented loss of control over the aircraft.
Design improvement (21 %)	State-of-the-Art equipment was not installed in the aircraft.
Maintenance (19 %)	Maintenance procedures must be improved to a degree that in-flight emergencies occur less frequently.
Captain exercise of authority (16 %)	Timely intervention by the captain due to incorrect crew behaviour was lacking. A professional working environment was not present.
Approach path stability (15 %)	Unstabilized approach: configuration, speed, height, flight path
Go-around decision (14 %)	Improved go-around training under more difficult operational conditions in the simulator. Ensure that go-around decisions are made on the basis of published safety recommendations.
ATC (13 %)	Improve ATC hardware, controller performance and management.
Eliminate runway hazards (11 %)	Improve methods of avoiding the risk of collision on the runway. Improve taxiway and runway lighting. Improve holding point markings. Improve runway surface grooves for better drainage.
FO crosscheck—performance as Pilot Not Flying (11 %) and Captain's crosscheck—performance as Pilot not flying (3 %)	Monitor and immediately correct errors committed by the Pilot flying in an independent, critical, competent manner.
Weather information and accuracy (8 %)	Precise weather predictions should be made available.
Response to GPWS (7 %)	Improve training of Terrain avoidance procedures.

(continued)

Table 1.9 (continued)

Strategy	Explanation
Airport Services (6 %)	The removal of snow, ice and foreign objects should be improved. The deterrence of birds in the vicinity of the airport should be improved. Improved deicing capabilities should be provided.
Pilot Flying awareness and attention (6 %)	Avoid distractions. Maintain a Sterile Cockpit. avoid complacency and inattentiveness.
ATC/Crew communications (5 %)	Actively check all data, clearances and confirmations. Use standard R/T.
Pilot experience in aircraft type (5 %)	Do not place inexperienced pilots together in the cockpit. Improve the training of inexperienced pilots.
Pilot Flying communication or action (5 %) and Pilot not flying communication or action (5 %)	Intervene immediately, clearly and intelligibly in the event of deviations from the standard. The competence, attentiveness and commitment needed for this should be trained.
Crew fatigue (4 %)	Improve fatigue management.
Fire and rescue services (4 %)	Improve alarm response times and Crash/Fire/Rescue services. (This will not prevent an accident but will help to limit its consequences.)
Training for abnormal condition (3 %)	Place a greater emphasis on accident-oriented abnormal training. The main areas of focus should be dealt with to a greater degree.
Availability of approach aids (2 %)	Install more precise approach aids.
Manufacturing process (2 %)	Improve component reliability in the production process.
Performance data (2 %)	Prevent error potential in performance calculations through improved procedures.
Pilot incapacitation (2 %)	Train for (medically-, physiologically-, mentally-related) incapacitation recognition and treatment.
Use of all available approach aids (2 %)	Example: Use a functioning ILS even during visual approaches.
Weight and balance control (2 %)	Develop SOPs that prevent the inadvertent use of incorrect takeoff weights.

1.4.4 Conclusion

Empirical findings from the studies identify three ways to approach more effective accident prevention:
1. More stringent application of SOPs
2. Improved CRM
3. Improved basic flying

The requirement for more training of the obligatory "Abnormal Procedures" does not appear on this list. Yet, because the majority of accidents occurred during "Normal Operations", significantly greater emphasis should be placed on this key training aspect.

1.5 Consequences

1.5.1 Flight Operations

- Pilot selection and training should be optimized to the desired level of flight operation safety.
- With a change of employer, pilots should receive training as to how their individual work habits must be adapted to conform to the level of safety demanded by the new operating environment.
- Airline management should be cognizant of the investment that must be made into safety so the company's long-term economical basis is not destroyed by a short-term profit motive.
- In corporate groups comprised of multiple airlines, similar and uniform selection and training standards should be pursued where possible for of all branch flight operations. This will help produce a consistent level of safety.
- SOPs are the main key to flight safety. They must be known and applied. "Need-to-know" content must be defined and thoroughly trained, both in theory as well as in practice. Flight operations management must ensure there are no SOPs that are not, or are not adequately being complied with. Flight operations must become active when SOPs are identified that are not being sufficiently complied with: They must be modified or substantiated in detail and handled with greater emphasis during "Recurrent Training". All multipliers, such as management pilots and trainers, should maintain a uniformly high standard when applying the SOPs. Grey areas should leave as little room for interpretation as possible and must be clarified at the highest level of the flight operations hierarchy. All pilots should be aware of the mechanisms specified further below, which can weaken the disciplined application of SOPs.
- Personalized FDM feedback is sensible under very strict conditions (data protection outside the disciplinary hierarchy with operating partner veto rights).
- Furthermore, pilots should have the possibility of submitting confidential safety reports to safety pilots without fear of disciplinary action or legal consequences.

Moreover, a culture should be established within the flight operation, in which these confidential reports are actually submitted in writing. According to well founded estimates, only about 1 % of all safety–critical incidents occurring within a large German airline are reported in this manner.

- First and foremost, new FOs are vulnerable to accident. The "initial training" received at an airline new to the FO should be so extensive that he is capable of recognizing and addressing preferably all of the captain's mistakes on their first flight alone together (not with an additional FO).
- CRM must be precisely defined and integrated into every training event by the flight operation. If CRM is assessed during training, then it can be more effectively developed by the individual. A prerequisite for the assessment would be the development of a flight operation-based CRM Assessment Policy that allows no room for arbitrary action on the part of individual trainers. This book provides guidance to this end in the sections that follow.
- "Basic flying" must be improved upon. Initial Training should include the safe mastery of all levels of automation, basic jet flying, exploration of aircraft performance limits and the training of monitoring skills. Especially for long-haul pilots, measures should be taken to effectively compensate for the low "stick-time" common to this type of operation. An increased emphasis on basic flying training in the simulator can be implemented, as well the requirement for more hands-on flying using differing levels of automation in daily flight operations under precisely defined conditions of fatigue, weather and traffic density. Northwest Airlines defined these conditions in its Flight Operations Handbook, thereby encouraging their pilots to fly at reduced levels of automation (Landry 2006).

1.5.2 Individually "Sufficient" Safety vs. Objectively Necessary Safety, Part II

Numerous individual consequences were listed above, particularly within the context of the Lufthansa study. For this reason, this section will remain general and concern itself for the most part with the difficulties encountered when trying to comply with SOPs as safety rules.

An individual pilot in a career encompassing around 20,000 flight hours will most likely never be involved in an accident.

Because nothing really serious happens to him month for month, it is possible that he will intuitively or unconsciously call these safety rules into question. A certain degree of looseness can develop over time that may not directly damage the individual, but can lead to a significant safety risk on the whole.

Development of this behaviour—analogous to road traffic—is normal for the individual, nevertheless inappropriate. In the event of an accident, the resulting consequences under certain circumstances could be catastrophic for the flight operation and, with it, the entire pilot corps.

The *objectively necessary level of safety* conveyed during initial training with the airline tends to degenerate to an *individually sufficient level of safety*.

Pilots seem to be in a perpetual state of dilemma: on the one hand, they must painstakingly comply with the SOPs, yet, on the other, they should flexibly call them into question if, by ignoring them, an apparent greater level of safety may result. Flexibility is called for especially then, when fuel is running short, for example, or a passenger requires urgent medical attention, or smoke or fire is detected, etc.

These are incidents where a deviation from a standard procedure could possibly increase safety. Such deviations should and can actually be limited to only a few specific cases.

Purely operational reasons do not justify deviations from the SOPs. If an approach is progressing too high or too fast, the SOPs are the lifeline that differentiates between an acceptable and an unacceptable risk. When an SOP is violated, the borderline between objectively necessary and individually sufficient safety is crossed.

In a further study conducted by Lufthansa, many pilots remarked that they are oftentimes forced to deviate from an SOP in daily operations due to ATC requirements. These deviations are seen as unavoidable in day-to-day flight operations. The level of safety that remains is therefore considered to be sufficient. The study expressly point out that this is very critical. Everyone who knowingly deviates from a rule does so for the most part in the assumed belief that they are acting safely. But risks can not be minimized to the necessary degree in this manner (Lufthansa 2009).

The following example depicts the increased risk of a runway excursion (Landing Overrun) associated with SOP deviations during unstabilized approaches. Hereunto, the Dutch National Aerospace Laboratory (NLR) analysed this type of accident (Van Es 2005), whereby 400 landing overruns were recorded between 1970 and 2005. With around 800 million landings during this period, the risk equated to 0.5 accidents/million landings. 53 % of the overruns took place on "slippery" or "contaminated runways". The conditions listed below (see Table 1.10) contributed to the increased risk of a runway excursion by the factors listed:

Table 1.10 Landing overrun risk factors

Condition	Factor
Long landing	55
Excessive approach speed	38
Visual approach	27
High on approach	26
Non-precision approach	25
Slippery runway	12
Significant tailwind	5

Statistically, an accident rate of 0.5 accidents per million landings with a flight operation of 500,000 cycles per year would mean that a landing overrun will occur every two years: an objectively high risk.

An individual pilot flies about 10,000 cycles in his career. Statistically, only one out of 50 pilots will experience such an accident in their professional career: a potentially acceptable risk for the individual.

Several factors stand in the way of the disciplined adherence to the SOPs: it can be tedious and painstaking.

- For instance, all private discussions and distractions should be avoided at any time below an altitude of 10,000 feet. Is this being strictly adhered to at all times?
- Normal checklists are read many thousands of times over. It can easily happen that, if it becomes too routine, it will be read only superficially.
- The call sign should be given first during R/T readbacks. This, too, proves to be difficult in practice.

Laziness, nonchalance, excess routine and complacency are an ever-present enticement to infringe against those procedures of seemingly lessor importance.

First and foremost, operational decisions must always be based on risk avoidance and error minimization. Economic considerations (e.g. delay, fuel) must play only a subordinate role. Private interests (e.g. proceeding, shuttling) should have absolutely no influence on safety-relevant decisions. In the interest of critical self-assessment, every pilot should pose the question as to how a particular flight would have been judged within the context of an aircraft accident investigation. This question should provide a benchmark, against which professionalism can be measured.

Flight operations must also pose self-critical questions, such as whether the SOPs published in the handbooks are practical, are being effectively put into practice and are being correctly taught. Otherwise, the impression may be conveyed that SOPs serve merely the legal self-protection of the aircraft manufacturers or operators, among others.

In addition, there psychological findings that address the issue of a pilot's self-discipline These are dealt with briefly in the following paragraphs.

1.5.3 Acquired Carelessness

While flying, as in many areas of private and professional life, it can frequently be observed that pilots ignore existing risks and disregard elementary rules of safety. This behaviour can be explained using the "Theory of acquired carelessness" (Frey and Schulz-Hardt 1998).

In a state of carelessness, it may be assumed, for example, that an error won't have any substantially negative consequences. This is referred to as "acquired carelessness" because pilots are careless when they come "right out of flight school", but because they acquire carelessness as a result of certain learning experiences. These experiences can be catalogued:

Individual experience

Carelessness arises when dangerous behaviour (e.g. SOP violations) is repeated and remains without negative consequence. The more frequently and intensively this happens, the more rapidly carelessness emerges. Because, especially in the flying profession, many SOPs were developed as a result of only one accident, it can mean that violations against them may lead to a similar accident only after many years. Yet, prevention of the improbable is precisely the objective.

One's own positive experiences, such as those gained from poor weather approaches, can actually be dangerous. With each successful approach in convective weather, the positive outcome increases the probability that the same or even a greater risk will be taken the next time. For this reason, an experienced pilot may be inclined to underestimate an actual risk after the positive outcome of several high-risk approaches. This was identified as a contributing factor in the report following the Air France landing accident in Toronto in 2005 (TSB 2007).

Hedonism

Hedonism refers to the striving for, and the preserving of a positive state of being, whereby greater significance is place on the short-term rather than the long-term consequences. Carelessness can represent just such a positive state, because the exercise of care means an increase in near-term effort at first. Under certain circumstances, it may be more convenient to try to preserve an uncritical, uplifted disposition rather than to comply with an SOP.

80 % of all incidents took place at a time when at least one flight crew member interpreted the working environment as being disturbed. One-third of these cases involved this uplifted, excessively positive state (Lufthansa 1999). An effective means of combating this state is to comply with the Sterile Cockpit concept.

Imitation

Observing a person's apparent success despite their careless behaviour often leads to the imitation of that behaviour. The captain's (and each multiplier's) example in this regard is particularly important in order to keep a "caution is cowardice" attitude from developing.

Control illusion

People are inclined to over estimate their own degree of influence. The illusion that "I have everything under control" facilitates risky behaviour—even when risks are perceived.

Unrealistic optimism

Although pilots are aware of the origin and principle importance of the SOPs with respect to hazard avoidance, they may be persuaded that they are not personally at risk: "It won't happen to me!"

Fatalism

A fatalistic attitude serves to impede one from changing his personal behaviour despite the threat of danger. It encompasses the mindset: "There are so many procedures, we can't know them all anyway, let alone comply with them all the time—so what's the use of learning them in the first place?"

1.6 CRM, Human Factors and Non-Technical Skills

Many pilots are sceptical when it comes to CRM, possibly having been influenced by a bad experience. This is comprehensible insofar as the physiological and psychological fundamentals of CRM, as well as the detailed safety-relevant behavioural patterns desired, are oftentimes insufficiently defined or disclosed. Therefore, it has been the aspiration of the Vereinigung Cockpit (VC/German Airline Pilots Association) to make CRM as practice-oriented and efficient as possible.

VC sees opportunities for improvement in systematic training, including time spent in simulators and aircraft outside the typical seminar environment.

The numbers referenced earlier speak out clearly for improved CRM training:

- 90 % of all incidents: available facts are not taken into consideration.
- 80 % of all incidents take place in conjunction with a disturbed working environment.
- 80 % of all accidents reveal deficiencies in the leadership of and collaboration between the crew.
- 70 % of all accidents occur following incorrect decisions or a failure to make decisions.
- 53 % of all incidents reveal communication problems.
- 30 % of all accidents are based on inaccurate situational awareness.
- 25 % of all accidents reveal symptoms of excess stress.
- 16 % of all accidents can be prevented through the effective use of the captain's authority.
- 14 % of all accidents can be prevented through timely go-around decisions.
- 12 % of all incidents: „the captain goes it alone."
- 4 to 7 % of all accidents happen to fatigued crews.

On the basis of these statistics, it is obvious that a large percentage of the accidents or incidents could have been prevented through effective CRM. To accomplish this, an integral training concept and the consistent implementation on the part of each pilot is essential.

A comprehensive CRM training concept will be presented later in the text comprised of the following content:

- Basic principles of information processing
- Fundamental approaches to dealing with errors
- Communication
- Stress management
- Decision making
- Leadership and team behaviour

> Only very few employees at the highest levels of the hierarchy can cause so much damage to an airline like a pilot can.

- Management of fatigue and attentiveness
- Implementation in the training
- Recommendations for an assessment policy

References

Airbus (2009) Flight safety overview, presentations at the German flight safety forum 2008

BFU (Bundesanstalt Für Flugunfalluntersuchung) (2008) (German Federal Bureau of Aircraft Accident Investigation) annual report 2007, Braunschweig

Boeing (1993) Accident prevention strategies 1982–1991. Boeing Commercial Airplanes, Seattle

Boeing (2009) Statistical summary of commercial jet airplane accidents, worldwide operations, 1959–2008. Boeing Commercial Airplanes, Seattle

Boeing (2011) Statistical summary of commercial jet airplane accidents, worldwide operations, 1959–2010. Boeing Commercial Airplanes, Seattle

Flight Safety Foundation (2009) http://www.flightsafety.org/gap.html. Accessed 10 Sept 2009

Flouris T (2006) San Jose State University, Lecture at the flight safety foundation EASS, Athens

Frey D, Schulz-Hardt S (1998) Erlernte Sorglosigkeit, Psychologie Heute, March 1998 edition

IATA (2009) IATA safety report 2008, Montreal

Landry D (2006) Lecture at the flight safety foundation EASS, Paris

Lufthansa (1999) Cockpit safety survey. Dept. FRA CF, Frankfurt a. M

Lufthansa (2009) Commercial aviation safety survey. Dept. FRA CF, Frankfurt a. M

National Transportation and Safety Board (NTSB) (1994) Safety Study NTSB/SS-94/01, Washington

National Civil Aviation Review Commission (1998) A safe flight into the next millennium. In: Flight Safety Foundation, Flight Safety Digest, Jan 1998

Reason J. (1991) Identifying the latent causes of aircraft accidents before and after the event. Lecture at the 22nd annual seminar » The international society of air safety investigators « , Canberra

Simon J, Mitchell LeM (2009) Cranfield University, Presentation at the ESASI Seminar 2009

Transportation Safety Board (TSB) (2007) Aviation investigation report A05H0002. Minister of Public Works and Government Services, Canada

Van Es G. (2005) Landing Overrun Study, NLR, Lecture at the Flight Safety Foundation EASS, Moscow

2 Information Assimilation and Processing

Gerhard Fahnenbruck

2.1 Introduction

Try, if you will, to teach a computer how to play tennis. It would be impossible, of course. Yet, the proposition alone points to just what outstanding levels of performance a human being is capable of achieving. A person's intelligence, his sensory skills and his motor skills are enormous.

It was through the human capacity to work with tools that he ultimately learned how to build aircraft, with which acceleration forces and speeds have been reached that man, himself, wasn't "built" for. Naturally, because this development has taken place over just a few generations, it has not been possible for him to genetically adapt, either.

This chapter will initially describe the mechanisms enhancing man's enormous capacity, what his limits are and what problems he must deal with, especially with regard to flying.

The section addressing "Information assimilation" describes the physiological principles of sight and hearing—particularly important senses for flying an aircraft.

In order to be able to process information once it has been taken in, or assimilated, it must be compared with information already stored in memory. Old information must be modified and new information added. This process requires various types of memory, as described in the section on "Information processing", in which their potential and their limitations are explained.

Two well-established models dealing with the processing of information stored in memory have been proposed in psychological literature. Both encompass only one part of "reality" yet, notwithstanding, are capable of describing the problems

G. Fahnenbruck (✉)
Main Airport Center, Vereinigung Cockpit e. V, Unterschweinstiege 10,
60549 Frankfurt, Deutschland
e-mail: gerhard.fahnenbruck@human-factor.biz

H.-J. Ebermann and J. Scheiderer (eds.), *Human Factors on the Flight Deck*,
DOI: 10.1007/978-3-642-31733-0_2, © Springer-Verlag Berlin Heidelberg 2013

associated with information processing and decision making in complex situations (e.g. in an aircraft).

All explanations and examples herein have been chosen so as to find application in the commercial airline industry. They have been drawn from the literature referenced, yet have been modified in part to facilitate better comprehension.

Overall, the fundamentals of information assimilation and processing as presented in this chapter will apply to most chapters in this book, and particularly the chapters dealing with "Human error" and "Decision making", as well as those dealing with "Communication", "Leadership and team behaviour" and "Stress" in a broader sense.

2.2 Information Assimilation: The Human Senses

2.2.1 General Considerations

The most important senses for flying are those of sight and hearing. Virtually all relevant information is taken in via these two senses and they are the primary focus of this chapter for this reason. In addition, the sense of equilibrium must also be considered. This sense may play a subordinate role in the conduct of a flight, but the information it provides inside the aircraft is easily and oftentimes misinterpreted. Problems associated with it are addressed in this discussion, as well.

Naturally, there are other helpful senses available to the pilot. A pilot will sense the aroma of a meal if the aircraft is equipped with an on-board galley. He will sense odours coming from the air conditioning system or the smell of smoke. Yet, as a rule, these senses are not important for carrying out a routine flight. If they do take on importance at some point, however, they will function just as they do on the ground. Deceptions to the sense of smell have, to date, only very rarely resulted in a wrong decision being made during flight. Similarly, this also applies to the perception of pressure, pain and temperature. For this reason, we have abstained from going into detail about these senses in this chapter.

2.2.2 The Human Eye

Most pilots have above average eyesight when compared to the rest of the population. Nevertheless, there are deceptions to perception related to the function of the sight organ, against which even good eyes cannot defend.

The following section addresses the healthy functioning of the eye, along with possible optical illusions and their related consequences for flying activities.

Functional Principle

Humans perceive light through the eye (see Fig. 2.1). Light initially penetrates the cornea, then passes through the iris and is refracted by the lens. An ingenious interplay between the iris and the lens ensures that objects perceived are clearly

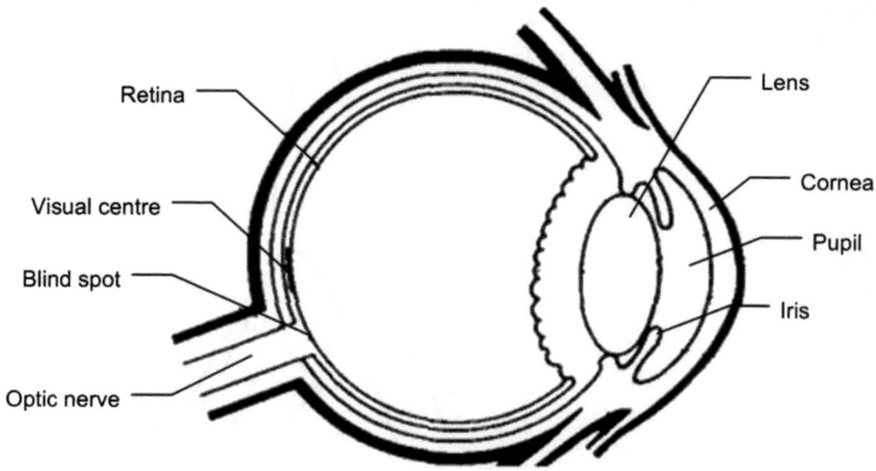

Fig. 2.1 The human eye

reproduced on the retina with optimal light intensity. This interplay takes place automatically for the most part and, therefore, can be influenced consciously only to a limited extent. For example purposes, just try to view an object with blurred focus. If you haven't attempted this before, you'll find it's possible only for short intervals and with a great deal of effort.

The retina is covered with four types of receptors capable of transforming light into electric signals. They differ in the wavelength of light that causes their maximum stimulation and in the intensity required for stimulation.

Three types are referred to as cones, which react particularly intensively to blue (445 nm), yellow-green (535 nm) and yellow–red (570 nm). They have a very dense arrangement with each being tied to a nerve fibre, enabling visual acuity. Cones require a relatively large amount of light in order to relay the respective colour impression to the brain, however, meaning that they function only in daylight or with sufficient lighting.

The fourth type of receptor is referred to as a rod, which reacts particularly intensively to green light (500 nm). As opposed to the cones, rods require very little light to stimulate a reaction and are arranged further apart from each other than are the cones. Moreover, a sole nerve cell may oftentimes be stimulated by multiple rods (see Fig. 2.2).

Cones (totalling approx. 6 million) and rods (approx. 120 million) are distributed non-uniformly over the retina. In fact, cones, alone, are located in the visual centre (approx. 400,000 per mm^2), while rods, alone, are located in the fringe area of the retina. This arrangement ensures that objects observed especially in natural light will be clearly distinguished with an accurate representation of colour.

The contrast-enhancing interconnection of the rods with one another ensures that the effects of brightness and darkness in the fringe area of the visual field are very clearly distinguished. In addition, rods also specialize in the perception of

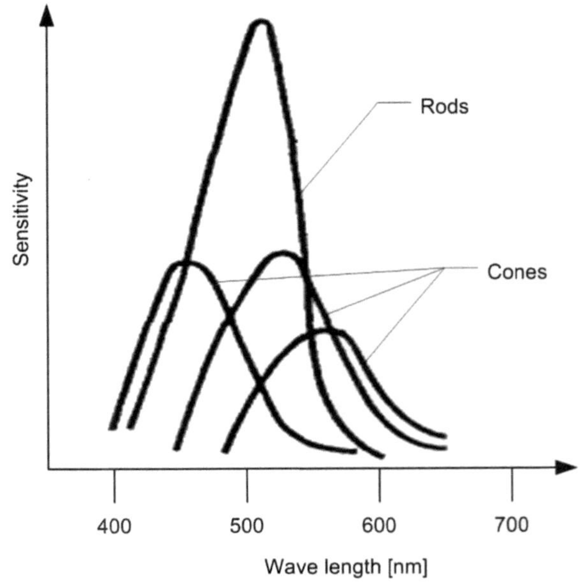

Fig. 2.2 Wavelength-related excitability per receptor-type

movement. If something moves in the fringe area of the visual field, it will be detected extremely quickly. Special processing within the brain is not required for this. On the other hand, sharp vision is not possible in the area. As viewed in evolutionary terms, this feature of the eye enables very quick reactions when being attacked from the side. In the event of an attack, however, sharp vision in the fringe areas would be counterproductive because the comparatively high volume of data would prolong the reaction time and consequently lower the chances of survival.

This retina "data" must be relayed to the brain via the nerve fibres for further processing. This is facilitated by the optic nerve, which emerges from the eye with a slight inward offset from the visual centre. This part of the eye is referred to as the blind spot because there are no receptors located there.

As long as sight is available in both eyes, these blind spots play an insignificant role because they are located at non-corresponding retinal points in the individual eyes. Corresponding retinal points in this context are understood to be points on the retina in the right and left eyes where one and the same perception is reproduced.

Humans also require the use of both eyes for distance perception. In order to reproduce objects on corresponding retinal points at distances less than 12 m, the eyes must be rotated towards each other at ever increasing angles as the object is positioned closer to the eyes. The brain "calculates" the distance between the object being observed and the eyes based on this rotation angle. At greater distances, the angular difference when focusing into infinity is so small that a precise estimation of distance with the given resolution of the eye is no longer possible.

In order to process the information taken in by the eyes, it must be relayed to the "proper" locations in the brain. But what are the proper locations? A person's

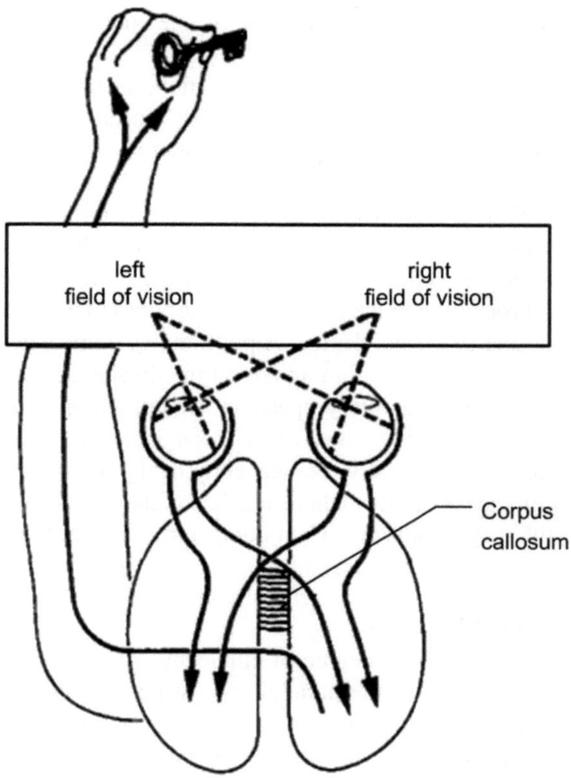

Fig. 2.3 Interconnection between the optic nerve and the brain

left arm, for example, is controlled by the right side of his brain, just as is the left leg. All motor functions and the overall sensory system (pressure, pain and temperature perceptions) on the left side of the body are governed by the right side of the brain. Correspondingly, the right side of the body is governed by the left side of the brain.

This "distribution of tasks" between the right and left sides of the brain with respect to the sensory and motor functions of the two body halves applies in principle to the eyes, as well (see Fig. 2.3). Everything seen on the left side (with either the right or the left eye) will be processed on the right side of the brain and vice versa.

Because of the division of the optic nerves into right and left visual fields, it is possible, for example, to leave control of the left arm through the eyes completely up to the right side of the brain. This "design" has a hidden benefit in that a transfer of data between the right and left sides of the brain is not necessary. Such a transfer of data would be possible via the so-called corpus callosum, which represents the only interconnection between the two halves of the cerebral cortex and, as such, represents a bottleneck factor, as well. If a related transfer of data were necessary, then control of the motor functions would become too time consuming and error-prone.

For processing optical information, an image of what is perceived is initially generated in the so-called "Area 17" in the rear section of the brain. Nerve cells capable or recognizing certain patterns, such as lines, circles or other simple figures, have access to this image. Additionally, there are nerve cells in the cerebral cortex that are specialised in higher-level processing (e.g. the reading of instrument indications). They access both the nerve cells in which the image is stored, as well as the nerve cells responsible for pattern recognition.

Processing in the brain takes place extremely rapidly because the brain works in parallel to a great degree. A pilot knows immediately when he observes a bar at a certain position within an arc on the display screen or instrument panel that he is dealing with a rotary speed indicator. All memorized information is available immediately once this observation is made. In contrast, a computer must be queried sequentially. The capacity of the human brain may be relatively low but, with the speeds that can be reached thanks to its parallelism during processing, it is still very impressive when compared to a computer.

Optical Illusions

The brain's high degree of efficiency when processing optical stimuli is possible only because of the parallel processing described above, as well as because of a systematic reduction of data. On the one hand, hardly any information from the periphery of the viewing area is consciously perceived. Processing of this data takes place only subconsciously, if at all.

On the other hand, the information consciously perceived is processed in a simplified manner. Psychologists were already trying to determine the rules applicable to this simplification process in the mid-19th Century.

Several examples of optical illusions presented in this section will demonstrate that not everything is as it appears to be when viewed. The optical illusions described herein have been taken from testing directives and, for this reason, may not always adapt to everyday application. The impact that even "minute" optical illusions can have during flight will be discussed in the next section (see Fig. 2.4).

In illusion (a), the right horizontal section appears longer than the left section. In illusion (b), the upper horizontal bar appears longer than the lower bar. Both can be described as illusions due to the addition of a third dimension. If, in the case of illusion (a), the horizontal lines are viewed as wall edges, then left section is seen as an advancing edge while the left is seen as a receding edge. The advancing edge must appear to be closer to the observer than it actually is because it jumps out of the image plane. Because it is apparently closer to the observer (and should actually be larger because of this proximity), it subjectively appears to be smaller.

The slanted lines in illusion (b) can be envisioned as being train tracks with the horizontal lines as railway ties. The upper tie is then farther away and, because it almost touches the tracks, must therefore be wider.

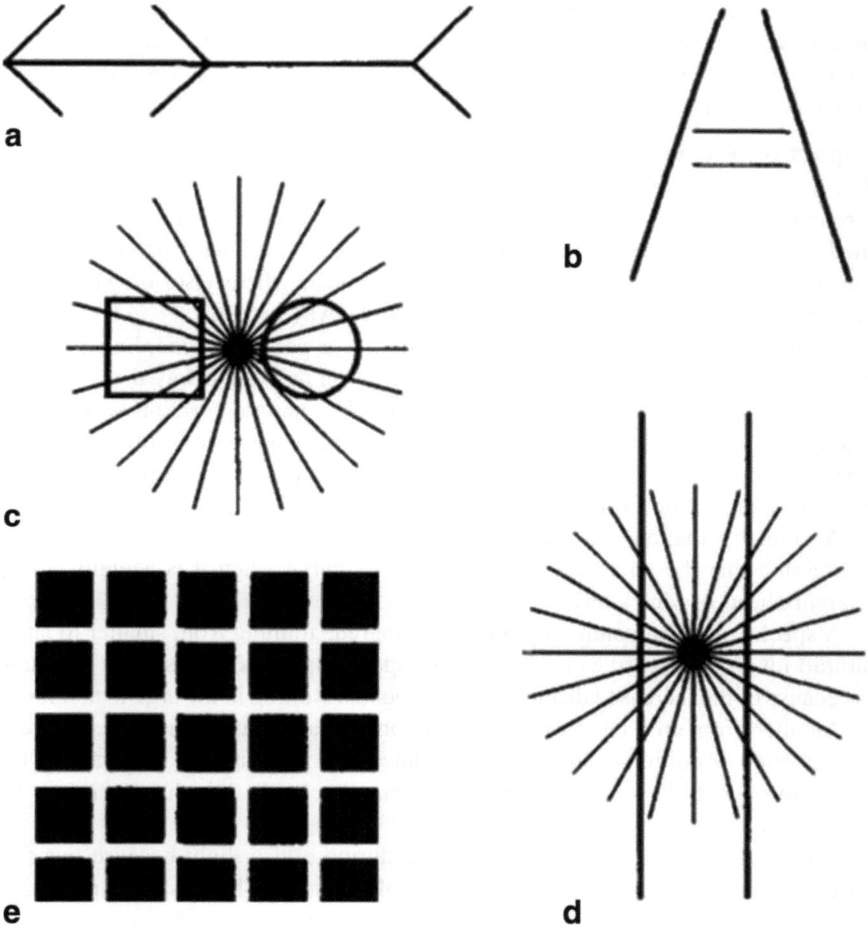

Fig. 2.4 Optical illusions. **a** Müller Lyer: The *right section* appears longer. **b** Ponzo: The *upper horizontal line* appears longer. **c** Orbison: The *square* and *circle* appear distorted. **d** Hering: The *parallel lines* appear warped. **e** Hering: The intersecting points of the *white bars* appear *grey*.

The deformation in illusion (c) or (d) is not due to the spatial interpretation of a two-dimensional figure, but rather due to varying information densities at different locations on the image. If numerous other lines are located between two lines at one location, then it follows that there must be more space between those lines at that location, otherwise the numerous lines could not fit between them. This applies to the circle or square, as well, which are segmented on one side by numerous lines. The respective segmented sides must be larger, because otherwise they could not have been divided so many times.

As seen in the contrast illusion (e), the ability of the eye to distinguish contrasts in a contrast-weak environment can, in a contrast-rich environment, lead to the illusion that the lighter sections of a picture appear darker.

Despite the obvious weaknesses of the sight organ, the eye is the dominant perceptive organ for determining the bodily position. Just about everyone has been on a train thinking it has departed the station, when actually it was a train on the neighboring platform that departed.

Impact on Flying

The processes described in the two previous sections play a role, virtually throughout the entire flight. An "estimation error" of merely half a degree during final approach can determine whether a landing will be hard or soft, whether touch-down will be at the 1,000 foot mark or somewhere else, or whether the aircraft will come to a stop by the end of the runway or not.

In the aircraft parking position, for instance, a passenger boarding bridge, a passenger bus or another vehicle in the vicinity can move in a way that creates an impression that the aircraft, itself, is moving. The natural reaction would be to immediately apply the brakes. It becomes critical when one knows about this "perceptual disorder" and, therefore, does not react. If the aircraft really is moving, then failing to respond can result in serious damage.

A similar situation can arise while taxiing. "Drifting snow" can create the impression that the aircraft is being taxied in a curve when it is actually rolling straight ahead or vice versa.

A special problem commonly related to upgrade training from smaller to larger aircraft involves excessive taxi speeds. The speed appears to be less than it actually is because of the increased distance to the ground, resulting in the urge to taxi faster.

Another problem related to directional control can occur during takeoff due to precipitation. Additional flight attitude-related illusions can be anticipated when flying through hilly and mountainous terrain or clouds, which can arouse the perception of a false horizon.

A collision hazard exists on takeoff and during cruise flight because our eyes specialise in perceiving motion in the peripheral areas. The risk of an in-flight collision is particularly great when the position of another aircraft does not change with relation to one's own position (related to direction), meaning a so-called fixed bearing exists (see Fig. 2.5).

In Fig. 2.5, the position of your aircraft relative to the positions of various other aircraft at the same points in time is depicted at times t-4 through t. A mid-air collision will occur at time-point t.

Common throughout is that a fixed bearing existed at each position. If such a situation exists, then a mid-air collision is likely. If this is not the case, then a collision is not possible as long as the direction and speed of the respective aircraft do not change.

Another difficulty exists particularly during cruise flight, when the eye automatically focuses on what is being seen through its lens. First of all, it should be noted that the eye accomplishes its task very effectively during daylight. But problems can occur when it gets dark or when the contrasts begin to weaken.

If the eye does not detect a contrast sharp enough to orient upon, most people will align their focus to a distance just short of one metre (about the distance to the

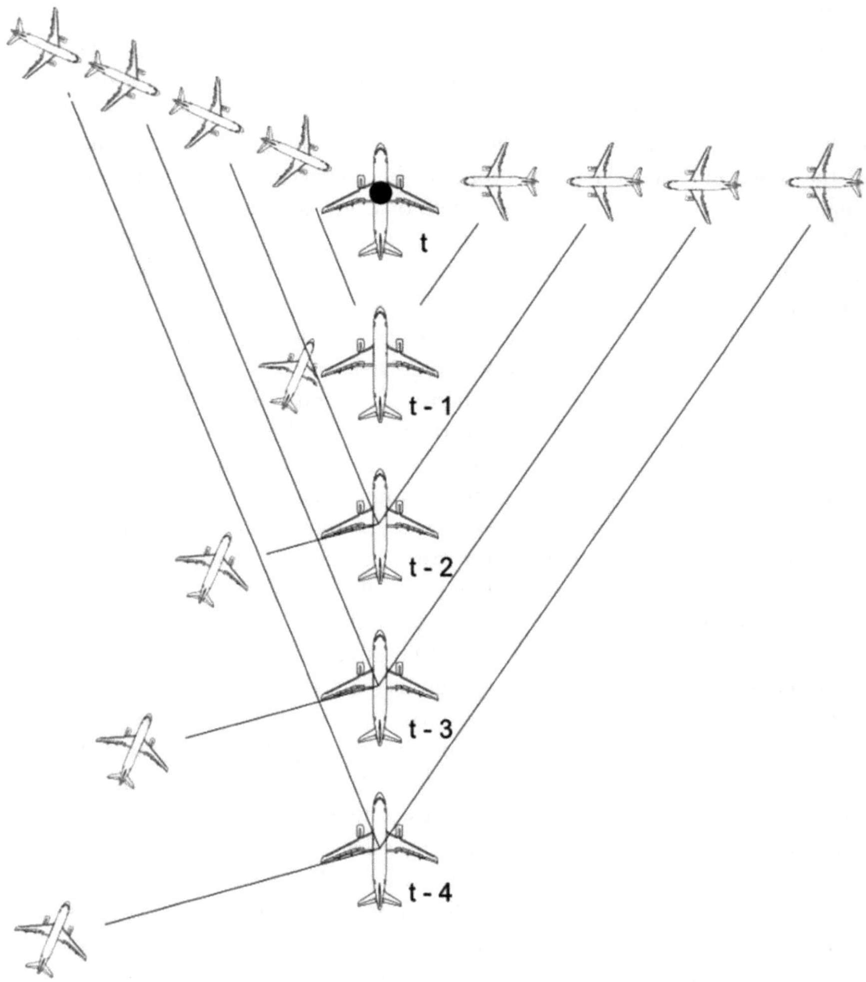

Fig. 2.5 Relative aircraft positions

cockpit windshield). When a pilot stares into the blue and frequently contrast-weak sky, his eyes, under some conditions, may focus more readily on the greater contrast provided by the potentially scratched, dirty or wet windscreens. The consequence can be another aircraft not being discovered due to an inadequate depth of field. This problem also occurs at night because the luminous intensity may not be sufficient enough to produce distinct contrasts. Moreover, if the blood circulation to the eyes is reduced (e.g. due to smoking in the cockpit), then its automatic focusing function will be additionally impaired. The aging process of the eyes, normally after 40 years of age, also plays a part.

The blind spot does not normally play a significant role. It becomes a factor only when sight in one of the eyes is impaired while the other is free. The chance of this happening in the cockpit increases with wider posts between the windshield panels.

The "interconnection" between the eyes normally presents no problem, meaning that everything seen in the right visual field will be stored in the left half of the brain and vice versa. Everyone having undergone upgrade training in the same aircraft type from copilot to captain or from captain to training captain, and who must now fly from the other seat, is familiar with the phenomenon that more time will be required to find a respective switch. It is possible that the corresponding information is stored on the "wrong" side of the brain. As soon as it is stored on both sides, however, the seat-change will no longer present a problem—as long as positions are switched on a regular basis.

The function and distribution of the rods and cones on the retina is optimized for use during daylight conditions. In contrast, difficulties can arise at night, when the cones are "blind" in the darkness. At the location where humans have the sharpest visual acuity during daylight conditions, namely at the visual centre, they don't see a thing there at night. The reflex action of looking towards a object when it moves or towards something of interest is, at the very least, useless at night, if not outright damaging. The act of looking to the side of the position one actually wants to observe is difficult to train and helpful only to a limited degree because focused vision and colour acuity are not possible.

Another factor to be considered at night is the so-called autokinetic effect where a single stationary light in an otherwise dark environment appears to move. There are essentially two factors that can cause this phenomenon. First, the eye requires the ear's vestibular system for stabilization of its viewing direction. This is not sufficiently precise enough to hold the eyes completely stable on its own, however. Secondly, the eyes move autonomously in order to keep from always stimulating the same receptors with the same information. This would lead very quickly to fatigue or even to the short-term "blindness" of the respective receptors. Both effects together will cause an object at rest to appear to be moving, especially at night. During the day, this effect is completely compensated for through experience. Every pilot knows that a runway does not move during approach, so this knowledge will stabilize his perception. At night, in contrast, it may not be possible under certain conditions to identify an unlit runway, so that the place where it is located appears to move.

Other problems can arise during approach, particularly in poor weather conditions.

It is common with *good visibility* to underestimate distances while they are overestimated with *poor visibility*. This is because, under poor conditions, objects being viewed are less easily identifiable, conveying the impression that they are farther away. The consequence for flight is that an approach may be flown at too low of an angle. A similar situation occurs at night when flying over unlit areas (water, desert, etc.). The pilot imagines himself to be higher than he actually is and will be tempted to fly lower in this case, as well.

Another possible interpretation for flying too low during poor weather conditions or at night assumes that, during a visual approach, the pilot normally

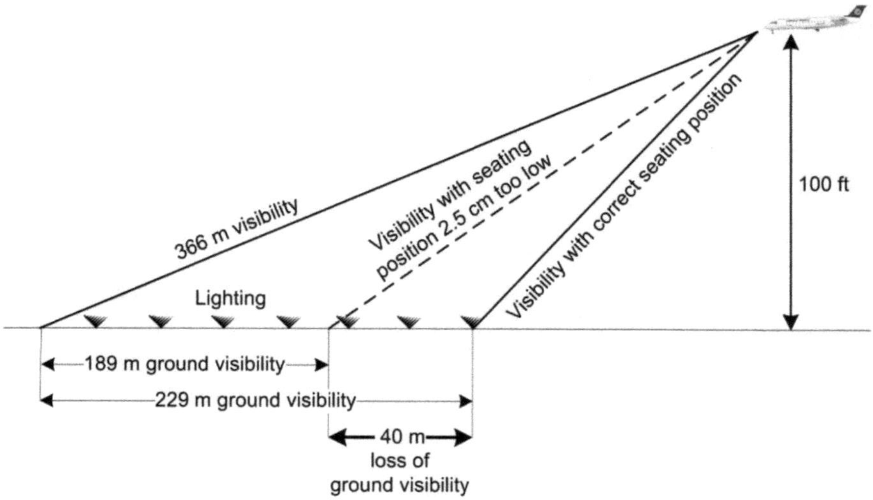

Fig. 2.6 Visibility loss due to incorrect seat position

maintains the angle between the natural horizon and the touchdown point, resulting in a constant approach angle. If the horizon is not visible, however, the eye will use the furthest point that may still be visible (e.g. the end of the runway) as a substitute horizon. Holding a constant angle between the substitute horizon and the touchdown point will automatically result in a downward arched approach profile.

Another problem can be related to the pilot's seat position during poor weather approaches, as illustrated in Fig. 2.6. If the pilot is seated just a bit too low, he will sacrifice a portion of his forward visibility. While the distinguishable lights located the farthest away will indeed be perceived, those lights located closer in will be blocked by the aircraft. This unnecessarily complicates the alignment with the runway.

Different approach angles due to varying approach speeds or flap settings can also have an effect similar to that of an incorrect seat position. The horizon will be located at a different position on the windscreen, meaning that the approach angle will be perceived differently, resulting in a potentially inaccurate flight path correction.

Runway gradients and sloping terrain can also lead to an incorrect estimation of altitude. Figure 2.7 illustrates four examples of possible altitude-related illusions. Mixed combinations of terrain and runway sloping are also possible.

If, in addition, the runway gradient is not constant but has "hills" or "valleys", then the length of the runway can be either under- or overestimated. Unusual runway dimensions (in length or width) can have a similar effect. Under certain circumstances, unnecessarily heavy braking may be used on a runway assumed to be too short, while insufficient braking may be used on a runway assumed to be too long. Both possibilities can have dangerous consequences.

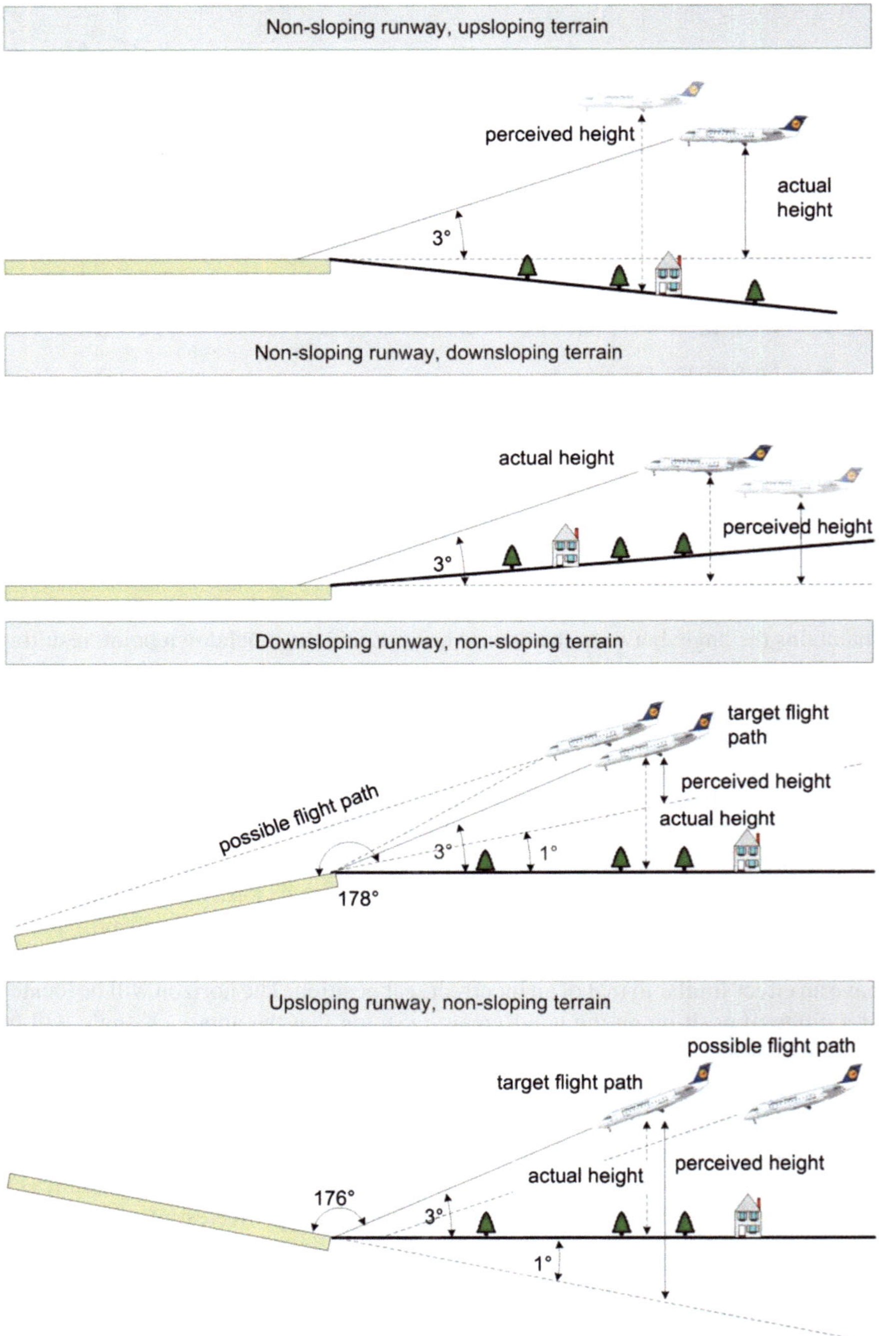

Fig. 2.7 Altitude illusion due to runway and terrain gradient

(No) Protection Against Optical Illusions

A method for protecting oneself against optical illusion does not exist. It is a phenomenon that everyone is susceptible to. It is human nature that the sight organ is subject to illusion. A training program applicable for all situations where illusions can occur cannot be developed. There are simply too many types.

Nevertheless, possibilities do exist for protecting oneself against the consequences of these illusions. First of all, illusions should be seen as being natural phenomena. Every person should become familiar with the situations conducive to prompting illusions and take advantage of all the materials available to help avoid them.

Cockpit windscreens should be cleaned prior to departure while seat positioning is clearly defined in transport aircraft. Using his knowledge of blind spots, a pilot can shift his position a bit to keep from impairing his forward visibility due to the cockpit windscreen posts. If it is known that aircraft appearing stationary in the sky can be dangerous, then a determined scan for their presence can be undertaken. The attitude that a plane will be spotted out of the corner of the eye, anyway, when the situation becomes critical is very risky. Just the opposite is the case. Aircraft will first become recognizable when they appear significantly larger, which, at the speeds commonly flown, will probably be too late.

A PAPI (or perhaps a VASIS) is frequently available as an aid to help stabilize an approach. As a rule, the ILS should be used when it is available, even during a visual approach. The crew member not flying should be tasked to take over altitude monitoring via DME or GPS and to call out any deviations, especially during non-precision approaches. Smoking prior to nighttime approaches to airports with poor lighting should be abstained from.

Moreover, authorities and employers should be tasked to establish procedures aimed at eliminating the consequences of the illusion phenomena.

2.2.3 The Human Ear

The human ear fulfils two functions. First, it facilitates hearing and, secondly, it serves the determination of position in space and, as such, facilitates balance. Both functions will be described in the following sections.

Because aviation-related disorders in the hearing process are relatively rare and can be traced back to either an overload of the working memory or the consequences of noise stressors (see the related chapter on "Stress"), this chapter will offer merely a short, simplified description of the process.

In contrast, the function of the ear as an equilibrium organ (vestibular system) for the flight activity has proven to be very prone to disorder. Similar to the eye, the ear was "conceived" for slow movements at ground level. The accelerations and the motion related to flight may be perceived, but may not always be correctly processed. The consequences range from discomfort and nausea to spatial disorientation. For this reason, we have dedicated an entire section to equilibrium organ-related disorders.

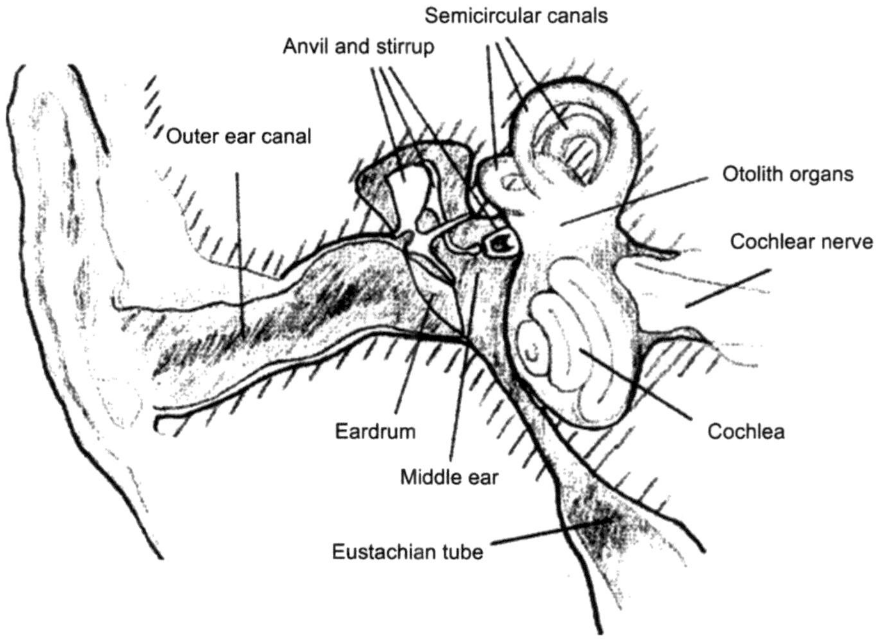

Fig. 2.8 The human ear

The Inner Ear

Sound waves penetrate the inner ear through the outer ear canal, the eardrum and the auditory ossicles (hammer, anvil and stirrup), causing the fluid in the cochlea to oscillate. These oscillations are picked up by extremely minute hair cell fibres and transferred via the auditory, or cochlear, nerve to the brain for processing (see Fig. 2.8).

Frequencies ranging from around 20 to 16,000 Hz are processed in the cochlea, whereby the human ear is particularly sensitive to the frequencies between approximately 1,000 and 4,000 Hz. This also happens to be the frequency range for human speech. Unlike the optical nerve, however, which is divided into the right and left visual fields, the auditory nerves in both ears provide both halves of the brain with their information.

A separate analysis of the information from the individual ears takes place merely with respect to directional hearing. For this, the time difference between a sound wave occurrence in the right and the left ear is measured and evaluated on the one hand, while a precise frequency analysis is carried out on the other. A noise that acts upon one ear directly and upon the other ear indirectly (passing by the facial area or the back of the head), or that simply contacts the two ears (auricles) from different angles, has its frequency distribution altered in a characteristic manner. An analysis of this distribution is performed in the brain with a bandpass filtering function that can distinguish clearly between tones as close together as 1,000 and 1,002 Hz, for example, thus enabling the direction of a noise source to be determined.

The Vestibular System

For determining position in space, the ear has three vertically stacked, so-called semicircular canals available for ascertaining rotational motion, as well as two areas comprised of so-called statoliths, which register the direction of the resulting forces imposed upon the body (see Fig. 2.8). The semicircular canals and the statoliths combine to make up the vestibular system.

The semicircular canals are filled with a fluid, whereby, if the head is swivelled in a direction on a plane with a semicircular canal, the fluid initially remains at rest. A relative motion then ensues between the fluid and the hair cells in the semicircular canal, with the stimulation of the hair cells being transmitted to the brain as a rotational motion in the respective plane. This principle works very nicely as long as the rotational motion does not continue for a prolonged period of time. There is absolutely no problem in rotating the body at a defined angle and being at risk of losing the sense of orientation. Similarly, a person will have little difficulty fixing his eyes on an object and simultaneously moving his head and/or body.

If the rotational motion continues for a prolonged period of time, however, then the fluid in the respective semicircular canal will come to a rest relative to that canal due to friction (similar to a glass being rotated over time, whose fluid takes on the rotational speed and comes to rest relative to the glass). The hair cells will then no longer report any motion in this plane. Once the rotation stops, a sense of rotating bodily motion will be induced in the opposite direction due to the inertia of the fluid, triggering corresponding reflexes. Everyone should know the phenomenon from their childhood, where someone is spun around a number of times, stopped and let loose. His body has the sensation of turning in the opposite direction, triggering related reflexes that will keep him from standing on his own legs for some time.

Statoliths also function very well on the ground. They make it possible to determine the direction of gravity, which is practically the only constant force exerted on a body under natural motion. Hair cells located in a calcite crystalline enriched gelatine layer stimulate their assigned sensory cells uniformly if the body is in a vertical attitude, or more or less one-sidedly if it is in a tilted attitude.

In an aircraft, however, statoliths interpret both the resultant of centrifugal force and of gravitational force as being gravity. The resulting misinterpretation can be compensated for only by means of the eye's monitoring function.

Disorders Related to the Equilibrium Organ in Flight

Disorders related to the sense of equilibrium in flight result from the design of the vestibular system. With an introduction of curved flight, the nerve fibres inside the semicircular canals sense the rotational motion that is actually taking place. Yet, the statoliths register a resultant vector that doesn't accurately reflect the rotational movement being carried out. Similar rotational motion on the ground would be accompanied by the risk of falling, so that any related reflexes must be suppressed.

If the rotational motion continues, the fluid in the semicircular canals will come to rest due to friction. The statoliths register the resultant vector so that the equilibrium organ assumes an unaccelerated horizontal flight attitude.

When recovering from curved flight, the equilibrium organ induces a sense of rotational motion in the opposite direction. This situation is accompanied by the related risk of a false reaction.

All intermediary states of a body in rotational motion and fluid rotating through the semicircular canals are possible. For this reason, an estimation of flight attitude based on the human sense of equilibrium is impossible. This is also the fundamental reason why VFR pilots have so many accidents when flying into IFR weather. They have greater trust in their sense of equilibrium than in their instruments.

Once an aircraft enters a spin condition, the same effects are greatly increased because all rotational axes are affected. A potential reaction when recovering from a spin could be to inadvertently enter a spin in the opposite direction.

Signals sent by the equilibrium organ are natural and should not be suppressed. The only thing a pilot can do to stave off their effects is to avoid head movements during instrument flight and to trust his instruments implicitly.

Even if a pilot has been trained to trust only what he can see, he will nevertheless be subject to the optical illusions described herein. In addition, there are also illusions associated with the interaction between the eyes and ears. Strobe lights, windshield wipers and propellers can induce an impression of motion that is not actually taking place or is taking place in a manner other than perceived. Because responses to such illusions are not uniform, however, it is advised in this case, as well, that the pilot should trust his instruments. If the pilot is unsure of himself, then he should switch on his autopilot (if available). In any case, it is easier to monitor the autopilot than to try to overcome a perceptual illusion while flying the aircraft at the same time.

2.3 Information Processing

The previous sections discussed how information is taken in by the eyes and the ears. Those sections described in part just how the first steps are taken in processing information and what system-related errors (can) occur thereby.

Once the effort is made to delve into a discussion of the higher, cognitive human processes, there is no way around dealing with the memory. Without memory, every piece of information taken in would be "new". Our condition would be similar to that of an infant's—a being without any prior experience whatsoever.

As long-term memory plays a particularly significant role in decision making on the flight deck, two related conceptions commonly held today will be described at length.

The potential for, and the limits of human information processing related to flight will be subsequently addressed. Limitations to this potential due of special situations can be found in practically every chapter in this book. This includes the chapters on Human error, Communication, Decision making and Stress.

2.3.1 The Human Memory

There are two different conception models commonly accepted today. The one differentiates between sensory memory, short-term memory and long-term memory. This relates closely to the experience that not all information is available all the time ("What was Peter's friend's name again?") and that not all available information can be taken in or processed simultaneously at all times. This multi-store memory model corresponds to the concept of information processing with the help of schemata or scripts.

The other memory-related conception model assumes one individual storage facility, yet with different conditions within this store.

While the multi-store model referred to above concentrates more intently on the aspect of the information being processed, the single-store model places the actual processing methods as the focus of consideration. Mental models for the processing of information fit more appropriately in the later conception about memory.

The question, as to which of these conception models is "appropriate" for aviation, is irrelevant. Of much greater significance is that the perspectives gained from either conception model can contribute to an understanding of the flight-related problems associated with information processing.

The Multi-Store Memory Model

The information taken in by the sensory organs initially accesses the so-called sensory memory. This has practically unlimited capacity with respect to volume. Yet, the information stored there exhibits an approximate half-life value of merely 100–150 ms. After this period, half of the information has already been lost.

The sensory memory is used to provide sufficient time for arriving information to be processed. This processing takes place in the short-term or working memory. Besides the information from the sensory memory, the working memory also accesses information from the long-term memory which has already been stored there. The working memory's storage capacity is relatively limited, however, so that if one's attention is drawn away from an object or a problem, the information in the working memory will be lost after only a few seconds. The working memory can process only 7 ± 2 units of information simultaneously. This processing principle applies to motoric areas, as well. Good jugglers can juggle up to a maximum of nine objects at the same time.

The respective capacity limit is dependent upon the person and his condition at the time.

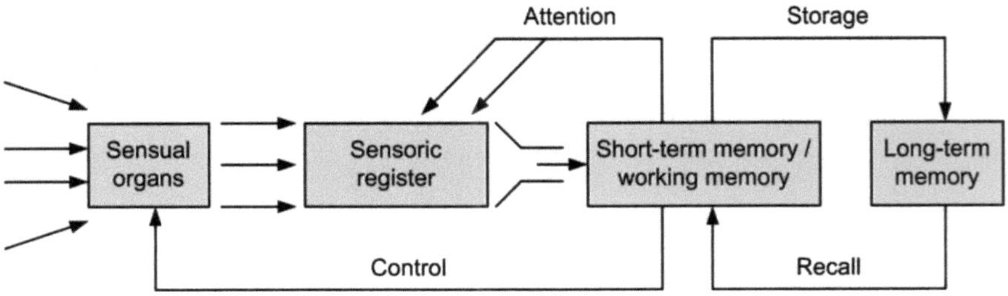

Fig. 2.9 Multi-store memory model

An overload of the working memory can occur if the approximate 7 units are exceeded. Additional units of information will be blocked and tunnel vision will ensue. Now, at the very latest, is the time to substantially reduce the load (with delay vectors, autopilot usage, etc.). Moreover, errors will also begin to occur within the approximate 7 units: A person who can memorize a 7-digit number without difficulty will most likely only be able to retain fewer than 7 digits when trying to memorize an 8-digit number.

Despite the limitation of approximately 7 units, the working memory has enormous capacity because it has access to the long-term memory (see Fig. 2.9). From this location, for example, experiences can be activated and combined with information within the working memory.

Information can enter into long-term memory from the working memory either consciously or unconsciously. Long-term memory is practically unlimited and comprises what the respective person knows about the world. Knowledge that hasn't been used for some time may not be easily recalled but, as a rule, is not lost. Even knowledge that a person can't remember despite lengthy deliberation can be recalled under hypnosis; an admittedly impractical technique for application in the daily flight routine, however.

Learning, with respect to this model, is the transfer of information into long-term memory by means of processing in the working memory. If the information to be learned remains active in the working memory long enough, it will then be permanently stored in long-term memory. This takes place, for example, through continuous repetition, through the linkage to a preferably large volume of content already stored in memory or through the active search for new information that can be linked to the content being learned.

The Single-Store Memory Model

The single-store memory model assumes merely one storage facility. All information coming in from the outside is either immediately processed in this memory or is lost. A personal long-term memory is not incorporated into this model. Moreover, the single-store model assumes that working memory is actually only one state of arousal within the memory.

If a portion of the newly arriving stimuli is recorded because of the state of arousal in a section of this memory, then it can be stored. The high degree of loss during information acquisition is explained in this model by a shallow depth of processing. Not all incoming information can be processed because the overall memory cannot be active at all times. For this reason, the unprocessed information is lost.

2.3.2 Schemata and Scripts

The terms schema and script play a large role in the theories dealing with knowledge stored in long-term memory. Both terms deal with knowledge correlations stored there. While schema describes knowledge through terms and definitions, script describes knowledge through action processes.

An example of schema can be found in the notion of a bird. A bird can fly, has wings, lays eggs, etc. A typical representative of this species would be a robin, while the features of a chicken or an ostrich diverge significantly from the schema drawn up by our conception of what a bird is.

Scripts represent stored information related to action processes. An example of a script can be found in a visit to a restaurant. The visit is comprised of: entering, being escorted to a table, reading the menu, ordering, eating, paying and leaving.

Schemata and scripts, each in their own right, can vary significantly. It's heavily dependent on the experience of the individual. This can mean, for example, that the script of being escorted to a table in a restaurant may have been stored differently by Americans than it might have been by Europeans. While almost all American restaurants will provide an escort to a table, this is rather the exception in Europe.

Schemata and scripts can greatly simplify human action. One will know how to behave if the corresponding script has been committed to memory. Yet, similar to how the simplification of information acquisition can cause problems in borderline situations, the simplification through the use of schemata or scripts can conceal the risk of falsification. If test persons are informed about, or see a film about a short story having to do with a restaurant visit, and are subsequently asked whether the customer paid or not, many of them will confirm that they heard or saw this take place, even if it didn't actually take place in the story.

Schemata and scripts play a similar role in flight. Power plant schema certainly appears differently to a jet pilot than it does to a sport pilot, while an engineer will see it differently than pilots altogether. Similarly, the script related to a flight sequence will differ significantly between an air transport pilot and an aerobatic pilot.

The problems that can arise when the schemata or scripts are stored incorrectly or when different members of the cockpit crew possess divergent schemata or scripts will be addressed in a later section.

2.3.3 Mental Models

If one assumes there is only one store for the entire memory, then the concept of a large network with many nodes, in which each node would compose a unit of knowledge, would seem likely. Each node can, in turn, then be linked to a differing number of other nodes. This network can be elevated through attentiveness. The respective information will then be available once a certain level has been reached. Likewise, new information will be introduced into this network when attention is paid to it. Such a network can be referred to as a "mental model".

Example: A colleague named Peter has a wife and three children. The last time I flew with him was on a trip to Rome two months ago. If I try to recall the weather conditions on that day, meaning to access a special node in my knowledge network, then I have the option of trying to do this directly. If that doesn't work, then I can attempt to remember through neighboring nodes. If one node in a network is elevated, then the neighboring nodes will also be elevated to a certain degree. I may recall that we had a slight slot delay, that Peter's two daughters are named Julia and Andrea, that Peter told me this during the flight to Rome and that we were looking forward to the pleasant day there!

This conception of memory as a network is generally helpful when trying to remember something not easily recalled. One simply attempts to elevate the network closest to the content trying to be remembered, then, at some point, the information will be activated to a degree that it can actually be recalled. This process requires some time, but functions very reliably.

2.3.4 Information Processing in Aviation

Several problem areas can be derived from the conception of memory and how it works that apply to the flying activity, as well.

Diverse aspects to consider can be drawn immediately from the working memory limitation of 7 ± 2 units.

First, burdens brought into the cockpit from the outside will limit the available working memory. Individual units may be "occupied" by a personal situation, such as home construction or some other issue, to a degree that only a residual number of units remain available for the actual flight operation (refer to the chapter on Stress).

Secondly, poor flight preparation is necessarily accompanied by smaller knowledge units. Under certain conditions, some information may have to be reworked during flight, which can significantly limit working memory capacity.

Thirdly, pilot training that is too accelerated or of insufficient quality can also result in the size of the units to be processed being smaller than their potential or less than what is needed. In this case, the working memory may be functioning at the extent of its capacity already during normal flight operations. Related errors during critical flight situations are that much more likely to occur.

The working memory limitations depicted above fundamentally reduce "situational awareness".

Fatigue, awkwardly designed cockpits, etc., contribute in part to a pilot not being totally aware of his situation, as well. Although cockpit safety can be enhanced by favourable ancillary conditions, such as through a reasonable presentation on the navigation display or with thorough briefing habits, the overall benefit to safety is nullified, however, if the pilot's capacity is limited because to the factors mentioned above.

It goes without saying that the responsibility for adequate flight preparation rests with the flight crew. It is also the pilot's responsibility to restrain from flying when a difficult situation may limit his flying fitness. What does not fall under his responsibility is the design of the flight deck and the establishment of training guidelines.

When crew members with different flying backgrounds occupy the same cockpit, there is always the risk and there is always the opportunity associated with their distinct insights into flight-related correlations. Scripts, schemata and mental models can diverge significantly from one another. One example might be the military pilot flying with a retrained flight engineer. Pilots coming from the business aviation environment will also possess different flying experiences than airline pilots.

A particular risk of misunderstanding always exists with such crew configurations. Yet, conversely, there also exists the chance that one crew member will more readily recognize an error made by the other crew member precisely because his correlated insight is different. It is possible to exploit the overall knowledge possessed by the crew through skilful communication, thereby realizing beneficial synergies. Helpful information related to this subject can be found in the chapters "Communication" and "Leadership and team behaviour". One such synergy, or complement to the different areas of knowledge, can be seen in the following example:

A captain, new to the aircraft type, is flying with a copilot who already has extensive experience in it. Even though the captain possesses the greater overall flight experience, the copilot may possess better experience when it comes to individual aspects of the operation. The captain should consciously put this to good use. Both crew members working together will conduct the flight more effectively than either one could do on their own.

For clarification: Despite the limitations that every pilot, every aircraft manufacturer, every employer and every aeronautical authority should be aware of, it is the human being, alone, who is sufficiently adaptable to, and capable of safely handling the complex environment found in today's cockpit.

Naturally, humans have their limitations; they make mistakes. They have not developed into beings capable of processing of large quantities of data in parallel. Every party involved should be reminded of this time and again in order to keep pilots from being put into situations that are conducive to error.

Naturally, humans have their limitations; they make mistakes. They have not developed into beings capable of processing of large quantities of data in parallel. Every party involved should be reminded of this time and again in order to keep pilots from being put into situations that are conducive to error.

Further Reading

Bösel R (1981) Physiologische Psychologie. De Gruyter, Berlin
Dorsch F (1982) Psychologisches Wörterbuch. Verlag Hans Huber, Bern
Hawkins FH (1993) Human Factors in Flight. Avebury Technical, Aldershot
O'Hare D, Roscoe S (1990) Flight deck performance, the human factor. Iowa State Univ. Press, Ames
Spada H (1990) Allgemeine psychologie. Verlag Hans Huber, Bern
Wiener EL, Nagel DC (1988) Human factors in aviation. Academic Press, San Diego

Human Error

3

Rolf Wiedemann

3.1 Introduction

Errare humanum est. (Cicero)	Piloti humani sunt. (VC)

Pilot error–human error–operating error–fatigue–these are all terms we hear about and read about over and over again in connection with hazardous events and even accidents. They all allude to the facts and may shed light on different aspects related to the cause but, at the same time, they insinuate that a person made a mistake while operating a machine.

As we know, the statistics ascribe a high percentage of civil aviation accidents to these failings (Lufthansa 1999).

Every accident is followed by lengthy efforts to identify the "guilty party" and the cause of his "wrongdoing". Once these are discovered, an attempt is be made to eliminate any of the weak areas identified, optimize the regulations and procedures, modify the operating environment and, sometimes, introduce technical changes and innovations, as well. Such an approach confines itself to the reactionary level instead of deliberately initiating a preventative process. This is especially true when, rather than a technical fault, an elusive "human factor" appears to be the cause. A technical fault is quite often easy to identify, comprehend and rectify. Human errors and their causes, on the other hand, are oftentimes difficult to comprehend.

Why is it so difficult to get to the root of these failings, with which we are reputed to have so much "experience"?

R. Wiedemann (✉)
Main Airport Center, Vereinigung Cockpit e.V, Unterschweinstiege 10,
60549 Frankfurt, Germany
e-mail: rolf@wiedemanns.com

H.-J. Ebermann and J. Scheiderer (eds.), *Human Factors on the Flight Deck*,
DOI: 10.1007/978-3-642-31733-0_3,

Everyone is familiar with the mix-ups, misunderstandings and operating errors that take place on a daily basis. Why was an incorrect "squawk" code dialed in, even though it was readback correctly? Why has no one noticed that the ILS has been set to the wrong frequency and now the airport appears from the right side, and not from the left side as expected? These and similar situations occur time and again, and we know that even with the utmost attention they cannot be completely eliminated.

Everyone should know that fatigue can easily lead to a degree of inattentiveness that can make it difficult to set control knobs, which are too small to begin with, or that can obstruct the view for what is essential with routine tasks, which there is an overabundance of.

Yet, as obvious as these examples are, the more unexplainable other occurrences seem to be. This, however, is only because an analytical investigation, analogous to those done for technical failures, is, for whatever reason, not performed.

We must be aware that not only do technical failures have clear-cut causes, but human error, as well, can be traced back to its root causes. It usually doesn't occur abruptly or coincidentally, but will be the result of a long chain of causal events. Certain schemata, situations and preconditions exist for it, too, that lead to the same error-prone situations time and again.

It is therefore all the more important to not just accept the error as being the end product of a series of coincidences. It should not suffice to merely search for the error in the system after the fact, but we must concern ourselves with the system in the error before it takes place. Only in this manner can a working environment be created that facilitates the early recognition of error-prone situations as they arise, as well as the appropriate response. Only with a better understanding of the origin of errors can they be specifically targeted, their frequency reduced and their damaging effects minimized.

Which factors are actually at work, where can the underlying problems be found and where should the focus be placed in order to create such an error-tolerant working environment? Only a careful analysis of the human working environment will help answer these questions.

The SHELL model introduced by Prof. Eldwin Edwards and Capt. Frank Hawkins (1987) describes the individual components of this working environment and their mutual interactions.

We will subsequently make an effort to classify errors in order to more closely address their causes. This knowledge will then be applied to the model of the working pilot prior to addressing the error chain and the pathway from the error to the accident.

The chapter concludes with a look at the potential for error prevention and error management. Both paths must be pursued, because optimal results can be expected only when they are dealt with together.

Because to err is human, the consequence of error must be limited through sensible workplace design.

Fig. 3.1 The human factor SHELL model

3.2 SHELL Model

The person occupies the centre position in the SHELL Model (ICAO 1989) (see Fig. 3.1).

Liveware I

The capabilities of this central component have been studied to a great extent. Size and shape, eating and sleeping habits, information assimilation and processing, expression potential, responsiveness and adaptability; these are all known. They are factors that form natural limits for the resilience of the overall system. The remaining components must be carefully adapted to these limits in order to avoid abrasive losses and error. These components are:

Liveware II

Colleagues on board the aircraft and on the ground, air traffic controllers, technicians and many others with whom he interacts; each within his own respective area. They supply him with information, issue instructions and provide support in the form of knowledge and cooperation. Naturally, they, too, are at the centre of their own SHELL systems and, just like him, they are human beings with their own weaknesses and limitations, albeit with their own unsurpassed capabilities and flexibilities, as well.

Hardware

The aircraft and its systems, displays and operating units, with which he works and which supply him with information and related faults to be dealt with. Of key importance is not only the technical potential of this equipment, however, but its ergonomics, as well, which plays a significant role in the usefulness, the practicability and, ultimately, the safety of the equipment.

Software

Guidelines and procedures, as well as all information such as flight plans, Notams, charts, etc., required for the work at hand. Comprehensible procedures, readable chart materials and Notams, as well as standardized and sensible work processes are good examples.

Environment

The person's surroundings, the geographic and climatic conditions, as well as other external components that influence flight operations. Similarly, other factors that can be influenced only to a limited degree, such as duty time and rest period regulations, or economical and even political stipulations.

It is particularly within this Hardware–Liveware relationship that aircraft and avionics manufacturers are challenged more than ever before to orient their products, not only on the technical options, but on the capabilities and the potential of the user, as well. This also applies to the Software–Liveware interface, where the sensible selection of, and the unambiguous presentation of available information are of particular importance.

In addition to the key interfacing links between the human being at the centre and the four factors encompassing him, there is a fifth interface that also should be added:

Liveware–Liveware (L–L)

This refers to the personal circumstances, the mental and physical condition, that can greatly influence job-related performance.

3.3 Classification of Errors

The classification of errors into groups and categories can be helpful for a better recognition of their backgrounds and causes. As in just about every case, a classification in this context is seldom clear because there are a multitude of possibilities with partially overlapping divisions, of which we will introduce only a few.

3.3.1 Error Forms and Error Types

While error forms describe the theoretical basis for errors, error types are oriented on the course of action taken in practice (Reason 1990). Error forms provide the theoretical background for this.

Error Forms

Error forms are errors whose causes are related to the manner, in which information is stored and processed in the brain. They are the result of the way our brain functions and are therefore the basis for every human error.

They are fundamentally based on cognitive mechanisms. The capacity for remembering and the processes for accessing and processing stored knowledge play a central role in this.

Reason refers primarily to two models for describing how our long-term memory functions: Similarity matching and frequency gambling.

When putting the individual pieces of information together into an overall picture, stored situation (schema) is initially called upon to provide a foundation, into which the greatest amount of available information most closely resembling the current situation fits (similarity matching).

When several compatible possibilities exist, the schema used most frequently to this point will be favoured (frequency gambling).

Example: A generator fails during descent. Without further information, a schema is activated containing a whole range of possible consequences and actions. Additional information (e.g. low oil pressure) invokes a completely different schema; engine failure.

Tversky and Kahnemann refer to these two mechanisms as the Rule of Availability and the Rule of Representativeness (Tversky and Kahnemann 1973, 1974).

They will be discussed in more detail in connection with the decision process in Sect. 3.4.2, Decision making. The term "schema" will be addressed in somewhat more detail in Sect. 3.4.1.

Error Types

Error types are the possible errors assigned to a phase of activity. An activity in this context is divided into three phases:

- Planning
- Processing
- Execution

Fig. 3.2 Basic error types (Reason 1991)

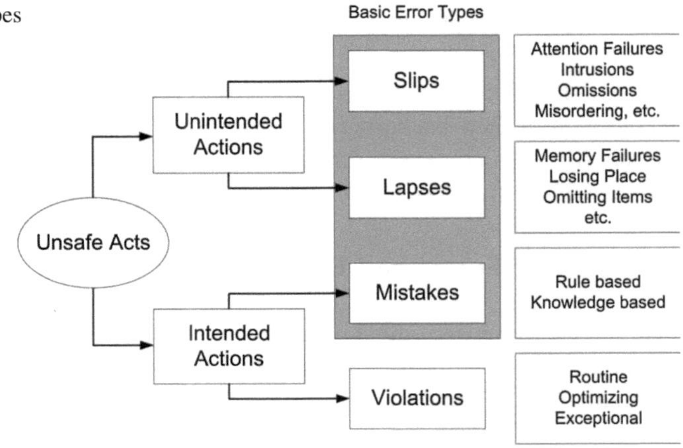

The "Unsafe acts" phrase used in Sect. 3.2 below connects the three fundamental error types, value-free, with the violations. It appears once again with the error chain in Sect. 3.5.1 (Fig. 3.2).

While lapses and slips "happen", mistakes are "made". Mistakes made intentionally are called violations.

Mistakes and Violations

Mistakes are made during the planning phase, perhaps due to erroneous planning data or incorrect conclusions. One refers to:

Rule-based mistakes, when the wrong rule is applied to a known situation, or when the right rule is applied incorrectly,

Knowledge-based mistakes, when a known situation is assessed incorrectly due to insufficient or wrong information (incorrect checklist, unknown airport).

Violations are comprised of all types of rule infringements (procedural shortcuts due to routine, well intentioned optimization, emergency authority). Violations are not basic error types, because their onset requires planning, processing and execution in each case.

Lapses and Slips

Errors during information processing: Lapses usually have something to do with the way humans assimilate and process information (forgetting, remembering incorrectly). The entire range of possible errors described under error forms can be related to this.

Errors during execution: Slips are behaviour patterns that are accessed at the wrong time (mix-ups, omissions, operating errors). These refer to the multifaceted errors attributed to acquired behaviourisms, the so-called motor programs, such as a new captain referring to himself as the co-pilot during the passenger announcement, or any slip of the tongue, among others.

3.3.2 Further Classifications

Active Failure and Latent Failure

This classification (Reason 1990) differentiates according to the temporal relationship between the activity and its effect.

An active failure usually has immediate negative consequences. It normally occurs during daily operations. "Gear up" instead of "Flaps up" during a touch-and-go is probably the best known example of this.

A latent failure is usually committed long before the actual accident. It is the result of a decision or an action, whose consequences remained undiscovered for some time. These failures are often caused by people who are far removed from the actual mission in terms of time and space. Examples of this may be found in management, the legislative process or operational procedures.

Commission, Omission, Substitution

This classification (Hawkins 1987) is broken down according to the fundamental type of failure. Commission is where an action is carried out that is not appropriate at the present point in time. Omission is where an appropriate action is forgotten, while substitution is where it is carried out on the wrong object.

Reversible, Irreversible

In this context, the classification is determined by the consequences. An error that can be undone and, for this reason, may not necessarily have serious repercussions, is referred to as being reversible (e.g. incorrect squawk code, frequency or flap setting). The consequences of an irreversible error, on the other hand, can no longer be influenced (fuel dumping).

Design Induced, Operator Induced

A failure at the Liveware–Hardware or Liveware–Software interface can result, for instance, from an operating system being inadequately adapted to the user (Hawkins 1987). This is primarily a problem of ergonomics and is therefore referred to as a "design induced" error. There are many examples of this, such as switches and levers being too closely positioned or too similar in appearance, or information being presented in a manner that can be easily misunderstood.

Similarly, the cause may be solely "operator induced" if dealing with simple operating errors not influenced from the outside.

Random, Systematic, Sporadic

With respect to their frequency and distribution, errors are distinguished as being

- *Random* haphazard error distribution
- *Systematic* orderly error distribution
- *Sporadic* intermittent error distribution

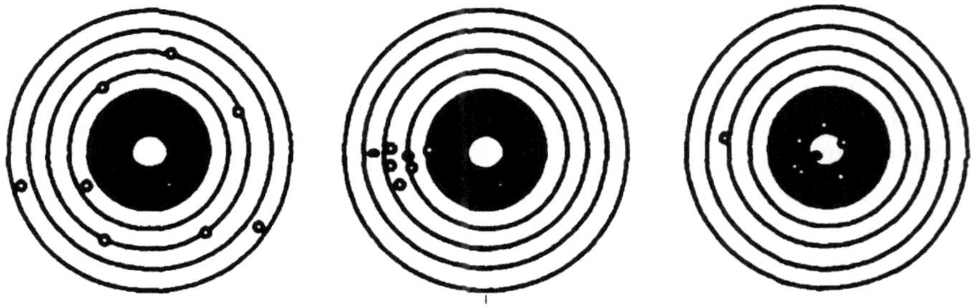

Fig. 3.3 Classification of error (Hawkins 1987)

With haphazard or arbitrary dispersion (random), the scope of possible errors is very broad, while with orderly dispersion it is very narrow. Sporadic errors are the most difficult to combat. Their emergence is not foreseeable and can have various causes (see Fig. 3.3).

3.4 Simplified Model of a Pilot at Work

The extremely simplified flowchart in Fig. 3.4 depicting work in the cockpit should make one thing particularly clear:

As with the SHELL Model, the potential for error exists at each and every link in the action chain, as well as at the junctions between them.

Some points will only be noted briefly here, as they are described in more detail in their respective chapters.

3.4.1 Information Assimilation and Information Processing

> We can only perceive that which we can conceive (Green et al. 1991).

Schemata (Mental Models)

Just as the receptiveness of the eye is limited to the range of visible light, our mind can perceive only those things that correspond to its conception of the world. Conversely, all perceptions are pressed into an existing model of the world, even if they don't actually fit.

Our mind's model of the world is comprised of a multitude of individual models, or so-called schemata (Reason 1990; Bartlett 1932). These have been stored in long-term memory from the earliest childhood on and are activated there through key stimuli.

Schemata reduce the effort of collecting information by providing ready-made mental models, within which only particular aspects will need to be modified or adapted.

The word "room", for example, calls up a schema already containing the basic characteristics of a room (four walls, door, windows, ceiling, etc.). This picture can now be filled in with the additional information provided. As already addressed in

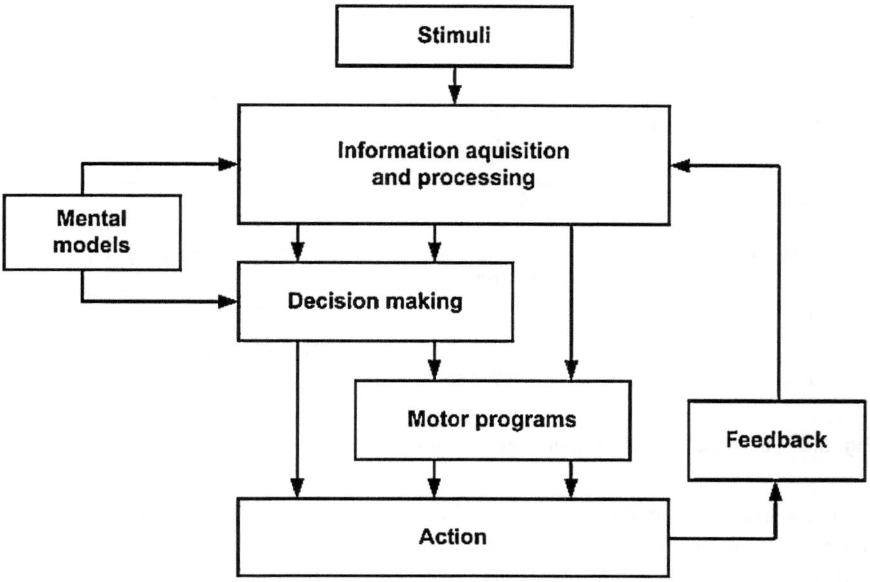

Fig. 3.4 Action diagram

Sect. 3.3.1, Error forms, the selection of schemata takes place by means of frequency gambling or similarity matching. The less familiar the situation, the lower the probability of encountering a valid schema.

When a compatible schema doesn't exist (unknown situation), a very elaborate and labour-intensive process begins that ultimately results in the formation of a new schema. Albeit, where time and decision pressures exist, optimal results will very rarely be realized.

The potential for error obviously exists where the schema contains information missing in the real world. We are very adept when it comes to introducing new data to a schema when it has been called up, yet it is very difficult to remove details from that schema or to distinguish between gathered and stored information in retrospect. Furthermore, once activated, a schema is very long-lived because confirmations are constantly being pursued while inconsistencies are ignored.

Information Processing by Humans

> Although the human is an exquisite processor of information by almost any measure, all of these means of acquiring information are subject to error. Thus, it is not only possible but likely that pilots will suffer lapses in their ability to maintain an adequate theory of the situation (Bohlman 1979).

The chart below (see Fig. 3.5) depicts one of four possible models that attempt to describe the working function of the human brain as it assimilates and processes information and should serve merely to provide an overview in this context.

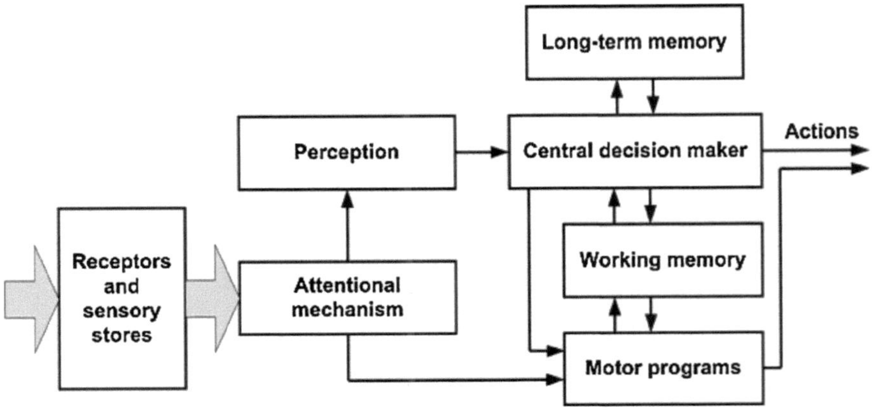

Fig. 3.5 Functional model of human information processing

These types of charts exist in a staggering degree of diversity and complexity depending on the respective phenomenon they attempt to illustrate. This model comprises a "central decision maker" (CDM) who executes the task at hand virtually in series. This explains the limited capacity of human beings for accommodating information very nicely.

When the CDM is working at the limits of his capacity, important information must be stored temporarily. Each of the senses (seeing, hearing, feeling, etc.) has its own small, short-term memory, albeit with greatly limited capacity. Because of this, a sensory stimulus (e.g. noise) may still be perceived (e.g. heard) under certain conditions even though it is no longer physically present. When the stimulus is then received by the CDM, it can be placed into temporary storage once again for final processing. This takes place in the so-called working memory or short-term memory, which, as we know all too well, has a very low capacity, with the lifespan of the information retained therein being very short.

Newer models also use a parallel method of processing information.

Errors During Information Assimilation

> Our direct senses are often compelling indicators of the state of the world, even when they are in error (Nagel 1988).

With respect to information assimilation in the cockpit, the following three senses play the greatest roles:

- Sight
- Hearing
- Sense of equilibrium (equilibrium organ and sense of force)

Related errors, misunderstandings and illusions can also be generated as a result of these. The causes for errors occurring during information assimilation can

Table 3.1 Sight

Error type	Example
Adaptation of the eye to brightness or distance	
False interpretation	Three pointer altimeter
Poor legibility	Instrument lighting
Mix-ups	Autopilot heading/speed
Unrecognizable failures	Missing flags
Illusions, disorientation	VOR inbound/outbound
Insufficient monitoring	Engine indications

Table 3.2 Hearing

Error type	Example
Communication	Readback, hearback
Aural warnings	Numbers, volume, differentiation
Noise and stress	Cockpit noise, communication

oftentimes be found at the Software–Hardware interface. False information or information provided at the wrong time cannot be assimilated either.

- Sight (see Table 3.1)
- Hearing (see Table 3.2)
  About 85 % of all errors originate within the verbal communication used for transmitting information (Nagel 1988).
  Our ability to communicate takes on overriding significance because we work with people in all technical areas (crews, air traffic control, handling, etc.). The problems associated with the L–L interface are described in more detail in the chapter on Communication.
- Sense of equilibrium and force
  Generally speaking today, a growing tendency can be seen in the volume of visual information that must be assimilated (FMS, EFIS, etc.). In order to avoid overloading this channel of acquisition, or rather to acquire an increased capacity for handling critical situations, a better distribution of the information, even to the other senses, would be desirable. At the same time, the tactile sense is oftentimes disregarded or its significance is underrated (moveable throttle, autopilot-control connection, interconnected controls, etc.).

Once the visual channel has reached its maximum receptive capacity in high workload situations, the processing of additional information will be possible only through other channels. One example would be the auto speedbrake.

The visual workload during a landing in critical weather conditions is extremely high. For this reason, it does make a difference whether one hears the function of an automatic system, perceives it through a correspondingly large movement in the peripheral visual field or must verify it through a focused glance at (and making inward note of) a display indication.

When the Autothrottle is activated, a thrust lever moving by itself provides the pilot with information about the thrust control function through the sense of force, thereby relieving his sense of sight.

Because every sense has its own small short-term memory, as already pointed out, such distribution will lead to an improvement in the overall capacity, as well.

3.4.2 Decision Making and Mental Models

The ability to make quick decisions in flight is of vital importance. Nevertheless, the mechanism used to make these decisions is no different than that used by other decision makers.

> Experts tend to make the same errors as do the rest of us under certain circumstances (Nisbett 1988).

The information gathered by the senses is processed into a mental mode by the brain, as already described.

Decision Making with Insufficient Data

The lesser the information provided by the senses, the more imprecise the corresponding model will naturally be. The mystery is that the brain will supplement the missing pieces of the mosaic. It uses the long-term memory as a "database" for this purpose by accessing the schemata already mentioned. Because of the inadequacies of this database, however, it is only natural that errors will occur. Furthermore, it is understandable that differences in the databases exist, meaning that different people exposed to the same situation may develop different models.

Unfortunately, we tend to hang on to our own models, being constantly on the lookout for new sensory data to support them. At the same time, we tend to initially suppress any evidence produced that might refute a model, rationalizing that it "doesn't fit into the picture", referring to our mental model.

This is one of the keys to understanding that crew behaviour, which, after the fact, is oftentimes incomprehensible.

> Once having made a decision, people persevere in that course of action, even when evidence is substantial that the decision was the improper one (Nagel 1988).

Decision Making by Rule of Thumb

Our brain makes things relatively simple at first glance. Research has shown that it does not process highly complex algorithmic solutions, but forms simple logical relationships. Decisions are based on so-called "heuristics" or, in plain English, rules of thumb (Nagel 1988).

Examples for how these rules of thumb work can be found in the functioning principles of our brain, as already mentioned in Sect. 3.3.1, Error forms.

- Rule of representativeness

Situation B follows situation A because that's how it worked the last time. Obviously, this method may have a certain statistical likelihood of success, but little more.

- Rule of availability

Long-term memory stores events and information (schemata) just like they would be stored in a file cabinet. The older the event, the further to the rear will it be filed with its activation being that much more complex.

When events and information are needed in order to construct a new mental model, the brain will favour the simplest path to more recent memories, even when experiences lying further in the past may be better suited to the current situation. Critical information must be "re-filed" to the front time and again in order to warrant optimal decision making. But even the best decisions can be wrong. Murphy's Law, scientifically anchored in the chaos theory, also applies to decisions once they are made.

One way out of the dead-end is to consciously turn away from certain patterns of behaviour. The key concept here is situational awareness, which will be addressed in more detail further on in the text. Within the context of a mental model, this means that support for the current model of the surrounding environment should not be pursued through new data but, just the opposite, data should be sought out that will refute it. Decisions should be subject to review time and again because:

> Most all of us are more confident in our decisions than we typically have any right to be (Bohlman 1979).

3.5 From a Simple Error to the Accident

3.5.1 The Error Chain

"One error comes seldom alone" and "one error, alone, doesn't cause an accident" are both commonly used expressions. In fact, it is very rare that an accident or incident can be traced back to one single causal error (Reason 1990, 1991).

A vast number of pre-conditions are required in a chain of events in order to generate the momentum needed to break through even the last line of defence.

James Reason describes the funnel-like course this process takes (see Fig. 3.6) from its origin to the final triggering event, after which we read about it in the newspapers.

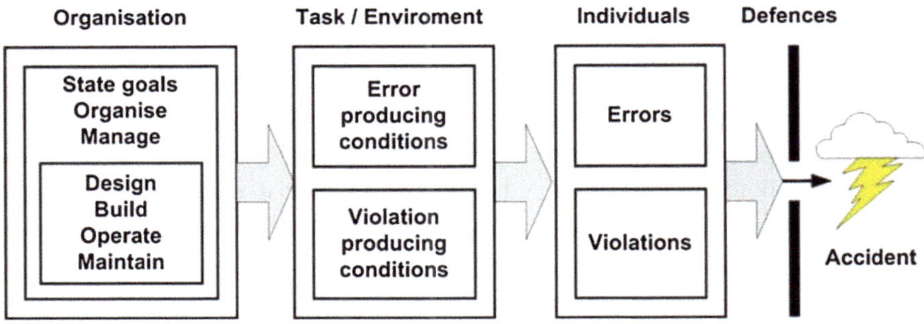

Fig. 3.6 Elements of an organizational accident (Reason 1991)

The first snowfall lays the foundation for a subsequent avalanche. Similarly, the error process can already begin at, and assume its overall breadth at the organisational level. Legal provisions such as flight duty time regulations and manufacturing specifications, aircraft manufacturer organisational structures, as well as their cultural differences, manufacturing philosophies and internal organisations, themselves, are all examples.

The task at hand and the work environment needed to accomplish it make up the next link in the chain. The environment must be very carefully adapted to both the task and to those who are assigned to carry it out in order to facilitate low work-related error rates.

The individual actor, with all his strengths and weaknesses, occupies the last position in the chain. As the last resort along the pathway to the accident, he is still in a position to compensate for built-in errors and system weaknesses or, through his own active failure, to trigger the avalanche referred to above.

Failures and weak points in the individual system groups are cumulative and weaken the tolerance for error throughout the overall system. These have been designated as latent failures (hidden weak points) in our classification.

3.5.2 Error Producing Conditions

Just how much influence his environment, latent failures and specific difficulties can have on an individual's frequency of error is depicted in Table 3.3[1] (Williams 1988).

The significance of the interfaces within the SHELL Model can also be clearly determined.

Most of these circumstances cannot be influenced by the pilot. Therefore, it is that much more important that they be regarded at the corresponding locations within the overall system.

---

[1] The factor indicates just how much the probability of error for a specific activity increases when the referenced condition exists.

Table 3.3 Increase in the probability of error

Condition	Factor	Source of error
Unfamiliarity with the task	17	Training, experience
Time shortage	11	System, environment
Poor signal/noise ratio	10	Environment, design
Poor human system interface	8	Design
Designer/user mismatch	8	Design
Irreversibility of errors	8	Design, system
Information overload	6	Design, system
Negative transfer between tasks	5	Design
Misperception of risk	4	Attitude, selection
Poor feedback from system	4	Design
Inexperience (not lack of training)	3	System, experience
Inadequate checking	3	System
Educational mismatch of person with task	2	Selection
Disturbed sleep patterns	1.6	Environment
Hostile environment	1.2	Environment
Monotony and boredom	1.1	Environment

3.5.3 Violation Producing Conditions

Violations are often the cause of accidents, although the corresponding relationships have not yet been well researched. The following table (Reason 1991) reveals that the overall system can also be mutually responsible for an individual's behaviour in this context; entirely in keeping with latent failures.

The external conditions are listed below, as well as the personal attitudes considered to be crucial for determining a violation, meaning a conscious or an accepted failure to fulfil or remain within applicable regulations or limits, in many of the accidents investigated.

- Manifest lack of organisational safety culture
- Conflict between management and staff
- Poor morale
- Poor supervision and checking
- Group norms condoning violations
- Misperception of hazards
- Perceived lack of management care and concern
- Little élan or pride in work
- A macho culture that encourages risk-taking
- Reliefs that bad outcomes won't happen

Table 3.4 Hazardous thoughts

Anti-authoritarian	SOPs are for others, but not for me
Impulsivity	Right away and make it quick
Invulnerability	That can't happen to me
Self-overestimation	I can do it
Resignation	What does it matter anyway?
Complacency	That's good enough
Exaggerated consideration	He must be right

- Low self-esteem
- Learned helplessness ("Who gives a damn anyway?")
- Perceived license to bend rules
- Ambiguous or apparently meaningless rules
- Age and Sex: young men violate

One example might be the various ways pilots from differing societies and different nationalities wear their uniforms more or less "correctly". Even though not necessarily accident-relevant, the disposition to commit such social "violations" can be traced back to several of the points above.

3.5.4 Hazardous Thoughts

A person contributes to the onset of error through his personal attitude about himself and his surroundings. His judgement and his conduct are influenced by five underlying attitudes in this model (Eberstein 1990), which are present and pronounced to varying degrees in everyone.

Too great an emphasis on individual components in this context transform the person, himself, into a latent failure. A simplified model defining the seven underlying attitudes is found in Table 3.4 (with a more elaborate discussion found in the chapter on Decision making under the section title "Hazardous attitudes"):

Every person can determine his own personal behaviour through simple tests, critical self-monitoring and feedback, and then influence it through conscious control efforts. Each of these attitudes is assigned an opposing thought (antidote), through which, when deliberately applied, the underlying attitudes that have become too pronounced can be defused (see Table 3.5).

3.6 Error Prevention and Error Management

> If you always do, what you always did, you will always get, what you always got (Wiener 1993).

Table 3.5 Hazardous thoughts—antidotes

Anti-authoritarian	Follow the SOPs
Impulsivity	Not so hasty, first take time to consider
Invulnerability	It can also happen to me
Self-overestimation	Don't take any risk
Resignation	I can make it happen
Complacency	Always strive for accuracy
Exaggerated consideration	Commensurate consideration

Accordingly, all the research into identifying the causes won't amount to anything as long as the results aren't introduced into the daily work environment. In order to learn from the mistakes of the past, not only the crews, but all the players in the commercial aviation industry, as well, must check their work and actions against new issues and adapt them to the new demands. The data collected from years of accident research and the resulting theories developed about the underlying processes must be converted into the safe, error-free execution of flight.

Hawkins (1987) divides the process into two steps:

- Error prevention (minimizing the occurrence of errors)
- Error treatment (reducing the consequences of errors)

The first priority is to take precautions in order to make the eventuality of an error as unlikely as at all possible. Then, because errors can never be discounted altogether, the second step must be to take counteractive measures to keep their consequences as slight as possible.

3.6.1 Error Prevention

SHELL Interface

A series of initiatives aimed at preventing errors has resulted from examining the SHELL Model and its critical points of abrasion. These have been discussed in part already and are therefore mentioned only briefly here. For the most part, they deal with requirements that must be fulfilled, though not by the crews in this case, however, but by the manufacturers, the responsible airline personnel, the authorities and the legislators.

Even if the user, himself, does not have a direct influence over aspects such as design-related deficiencies, himself, he should repeatedly draw attention to specific areas of weakness in order to affect long-term changes. Only in this manner can system-intrinsic weaknesses—latent failures—be eliminated over the course of time. The points listed below represent only a small selection of all the possible and relevant aspects related to optimizing the work environment, and virtually everyone could add to this list out of their own experience. With closer examination, however, it is that much more astonishing to discover that even to most » self-evident « rules are disregarded at times to some extent.

Liveware–Hardware, System Design

Adaptation of the machine to the man

- clearly readable displays presenting the proper scope of information in a well-arranged and easily interpretable manner
- standardized system of switches and operating controls that eliminates confusion and mix-ups

More and more, manufacturers are coming to the realization that the cockpit is not only a collection of system displays, but that its layout and configuration contribute significantly to the safe conduct of a flight. Admittedly, the spectrum of criteria that needs to be considered is very large. It starts with the size and shape of the switches, their positions, the structuring of the individual control units and displays, the presentation of information on the monitoring screens, just to name a few, and continues through to the size and lighting of the overall cockpit.

Even though scientific research into the design of displays was conducted as far back as 1968 (Roscoe), neither the manufacturers nor the national authorities have been able to agree upon unified set of international standards to date. Short-term economic considerations take priority time and again. At the same time, design engineers orient themselves more closely on the technical options and less on what is expedient for the person.

Liveware–Software, Software Design

Well-arranged presentation of all information

- clearly arranged organization of chart materials, Notams, aircraft documents and other sources of information
- sensible, comprehensible procedures
- well organized checklists

The software component is by far not as difficult to influence or to modify over the short-term as the hardware components might be; that is if one believes the promises made by numerous aircraft manufacturers. Actual practice shows, however, that modifications to software are at least as difficult to realize as the installation of new hardware. Intervention into highly complex, certified software is accompanied by a large financial commitment and always conceals the risk that new sources of error may be programmed in. Correspondingly, minor points of friction at the user interface won't be rectified when they can be remedied on the user-side much less expensively by issuing new operating instructions; meaning procedures.

An example of this might be the database for the Flight Management System, whose weaknesses pilots have long learned to live with. Because in this case as well: software and databases are created by humans, and humans simply do make mistakes. We encounter on a daily basis the portrayal of information that is either difficult to understand, unclear, incomplete, ambiguous or not suited to the situation.

The design of the checklist should not merely be oriented on whether all systems are correctly activated for the respective situation, but also on whether the workload is sensibly distributed, whether the peak workloads have been balanced out and whether it is conceived in such a manner that the crew members are encouraged work together.

Many research groups working in this field (communication design) have compiled and published related findings in recent years. NASA, for example, has released a collection of reports about accidents that can primarily be traced back to poor documentation. Accordingly, anyone who publishes information on paper or through other media in the aviation industry must, in fact, have sufficient resources available to ensure its optimal presentation with respect to human factors research.

It would considerably more convenient to be able to read a NOTAM date, for instance, not as a numerical series such as 0602080630, but rather in an easily discernable form such as 8 February 2006 beginning at 06:30. It not only simplifies the task, but it also helps prevent the errors associated with converting these numerical groupings.

Liveware–Environment, Environmental Shaping

We require an error-tolerant and stress-free working environment

- quiet and comfortable cockpits
- stress-free working relationships among crew members
- congenial working atmosphere

Unfortunately, we are not able to change many of the external circumstances. CAVOK conditions 365 days a year and a worldwide topography comparable to the north German plains would certainly be desirable. Other aspects, however, must be thought through anew and reassessed accordingly, as to whether they may be contributing to an unsafe, even error-conducive environment, incompatible to humans. One of the main human strengths is indeed his flexibility, but what about his physical and mental needs, without the fulfilment of which he won't remain motivated or effective for very long?

A working environment must be created that does not demand his flexibility from the outset, but is one in which he can work in a relaxed and concentrated manner, allowing him to call upon all his reserves in the event of an emergency. This includes reasonable statutory and operating regulations, optimized physical conditions with respect to noise, temperature and humidity, as well as a suitable "working atmosphere".

Liveware–Liveware, Inter-Human Relations

> Perhaps the best countermeasure is constant vigilance concerning the potential for errors in the entire process of communication, whether it is between pilot and controller or pilot and first officer (Bohlman 1979).

Influence over the interpersonal working atmosphere and optimization of the working relationship through:

- Crew resource management
- Crew coordination
- Crew performance instead of pilot performance

There is hardly a large company these days that can afford not to offer its employees advanced training opportunities, even when it is not necessarily related to their professional qualifications. This includes seminars related to self-assessment, working in groups, crew resource management and leadership. Besides mere technical knowledge and manual skills, it is also becoming more and more important to understand psychological correlations, as well as to adopt social and communicative skills. This subject will be discussed in more detail in the chapter on Communication.

Standard Operating Procedures

In today's two-man cockpits, the loss of a crew member means the loss of 50 % capacity with a loss of 100 % redundancy.

A crew member is considered to be lost when his mental model no longer agrees with reality. It has been shown in this model that the memory serves as a database.

If we use predominantly the same database, it also follows that any deviations in the mental models will be minor. As long as both pilots have the same SOP stored in their memories, the mental model, such as an ILS approach, will be generally in agreement without requiring a great deal of additional communicative effort.

SOPs can therefore be seen as a form of anticipated communication, or a type of elementary understanding about the working function presumed to be known from the outset.

When a deviation from a SOP is being planned, it necessarily results in the colleague being notified about this intent so he'll be able to adapt his behaviour (mental model) accordingly or assert his objections beforehand.

In the optimal case, the colleague should call for this information when he becomes aware that a deviation from his mental model is likely. It is the captain's responsibility to ensure a working atmosphere where the demand for this type of information is facilitated and supported at all times.

Selection, Training, Motivation

Selection

Darwin's theory is based on a form of natural selection. When choosing transport pilots, however, this type of selection should be avoided. Aptitude tests initially provide a financial benefit from the perspective of the one who will ultimately pay for the training, whether it is the airline or the applicant.

Oftentimes, it is not the occupational aptitude in the proper sense that is ascertained, however, but rather the statistical probability of successfully completing the pilot training program in order to reduce the financial risk and to ensure a high degree of productivity.

At the same time, just as the cockpit working environment has changed, so, too, is the demand profile imposed on today's pilots different from that of 20 years ago. In modern cockpits, it is no longer the manual skills that are of sole significance, but more often the management skills, the ability to work in teams and to coordinate processes, along with the flexibility needed to keep pace with technical innovations. The trend is moving away from "tradesman" and towards "manager with flying skills".

The selection of persons according to specific criteria will always result in a homogeneous professional group. Correspondingly, in the case of commercial airline pilots, this can lead to a lower susceptibility to error as long as team behaviour, communication skills and motivation are encouraged within this professional group.

In view of human factors research, it is particularly important to seek out people for this profession from the outset who demonstrate a high degree of conformance to the profile demands in order to prevent latent errors during the early stages of the career. The questions, as to which characteristics are testable and which criteria should be chosen in order to realize optimal results, have been the focus of many years of research and must be regularly adapted to the occupational profile, just as later commercial airline pilots must adapt to the changing demands of their profession.

It is in the interest of all involved parties, including the passengers, that the aviation authorities and legislators do their part to enhance aviation safety in this respect, assuming there is an interest to do so.

Training

> As with visual errors associated with the approach and landing, disorientation conditions are both compelling and avoidable if pilots are properly educated to the hazards and trained to either avoid the precursor conditions or to properly use flight instrumentation to provide guidance information (Bohlman 1979).

It is obvious that good flying skills and technical knowledge make up the cornerstone for safe flight operations. In the traditional sense, training means gaining proficiency over the aircraft's technical systems and its "standard abnormals".

It is evident from the increased activity in the field of crew coordination that other basic, more socially and psychologically oriented competencies, over and above the fundamental skills, are taking on greater significance and therefore should be integrated into the training program from the outset. In addition to this new content, other aspects of academic instruction and practical training should also be reconsidered and modified where needed.

In one study, it was verified that students were better able to successfully complete their training if they were given the opportunity to commit errors, recognize them on their own, and thereupon develop their own solutions (Frese and Altmann 1988). This is in contrast to the comparative group, in which errors were identified and doctrine was taught according to the classical model.

It is evidently needful to learn how to deal with one's own errors, as well, and not only with system-related faults or failures. This new form of error management plays an especially important role in connection with the LOFT program, in which students must accomplish a flight in real-time without help or suggestions from outside and, above all, without interruption.

Frese and Altmann (1988) describe this new approach in the following manner:

Change the attitude of trainees from » I mustn't make errors « to » let me see what I can learn from this error « .

This must also be put into practice by the trainers. In addition to the mere technical material being taught, didactical, pedagogical and psychological skills must also be conveyed.

Motivation

Many accidents blamed on human failure could neither be ascribed to poor design or unfavourable working conditions, nor to a lack of crew knowledge or skills. Oftentimes, the causes can't be accounted for at all.

In these cases, one encounters what psychologists relate to as motivation: Why does someone act the way they do? Motivation, in this sense of the word, denotes the difference between that which someone can do, and that which he does. There are various theories, as to which structures form the basis for this behaviour. It is generally agreed upon that there are several levels of desires and needs, whose craving for satisfaction differs in intensity. Maslow distinguishes between five levels (see Fig. 3.7):

The further down the unsatisfied need resides, the more expedient is its satisfaction; only after which, will the superordinate needs play a role. Differing behaviourisms can be influenced by various factors simultaneously and in opposing manners. A person's degree of satisfaction, and thereby his motivation, is greatly influenced by his corresponding field of work and working environment.

Relating this to the aviation industry means that an environment must be created that facilitates the highest degree of professional fulfilment and satisfactorily warrants the quality of the professional life. Three primary goals must be pursued in order to realize a safe and error-reducing work attitude:

- Prevention of complacency
- Attainment of a professional attitude
- Maintenance of discipline

These three points are the primary causes of those accidents that can be traced back to insufficient motivation.

Management is responsible for creating a work environment that facilitates motivated working. If they show up too late for work or leave before everybody else, it can be assumed that their employees with adopt the same attitude towards their work.

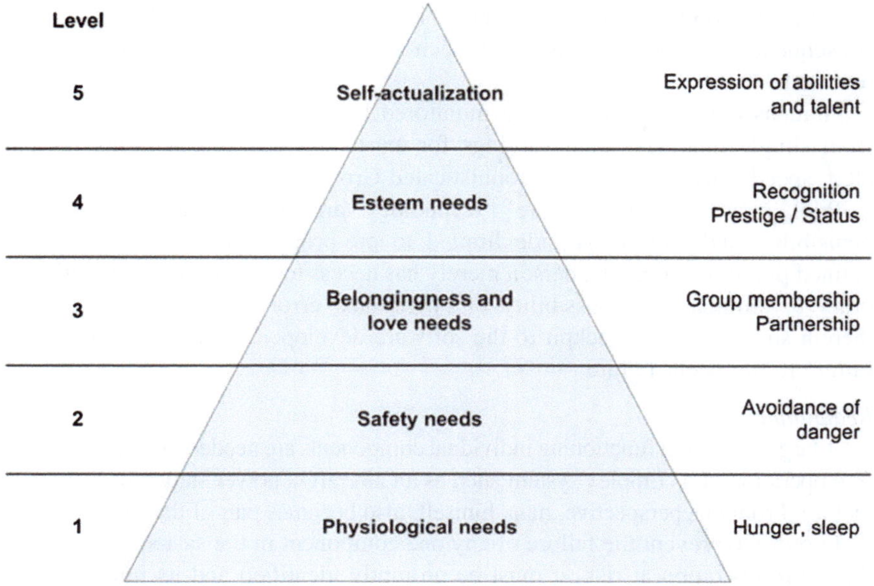

Fig. 3.7 Maslow's hierarchy of needs

Not only does the personal role model leave an impact, however, but recognition of the employee, his capabilities and his commitment, as well.

Other factors also play major roles, such as the acceptance of the profession by society, the rules governing leisure time and rest periods, the structuring of the duty schedule, the general corporate working climate and, not least, the income.

However, it would be improper to merely push off the responsibility and need for action onto "the others". Each person must examine himself, as to whether exorbitant expectations and a basic negative attitude have, themselves, become stumbling blocks along the way, ultimately leading to demotivation.

3.6.2 Error Management

System design

One option for limiting the consequences of error is to reverse them wherever possible. There is an array of examples, in which this concept finds application. For instance, when entering data into a display terminal by means of a so-called "scratch pad", all inputs can be checked initially before they are entered into the computer. Devices such as the "Gear Interlock System", or those that deactivate certain systems in the flight or ground mode, help prevent improper operations while facilitating the reversal of its consequences.

Wherever possible, systems whose improper operation could have grave consequences should be designed in such a manner that operator inputs can be reversed.

Humans are increasingly being monitored by technical systems. These range from simple interlocks and warnings for exceeding operational limits (altitude alert, speed clacker, etc.) to the sophisticated Ground Proximity Warning System.

Pilot inputs in "fly-by-wire" technology are electronically checked for plausibility and their magnitude limited to pre-programmed values. In certain defined phases of flight, the person merely has access to "permissible" functions in order to eliminate the possibility of operational error. The potential for error therein shifts from the cockpit to the software developer's office; from the active failure to the latent failure.

Redundancy

A large number of functioning individual components are needed to ensure the error-free operation of a complex system such as an aircraft, a power station or a chemical factory. From this perspective, man, himself, also becomes part of that system.

In order to prevent the failure of any one component in the system from putting the entire operation at risk, it must be promptly identified and its function taken over by other means.

Many different initiatives have been undertaken aimed at achieving this goal. "Fail Operational" and "Fail Passive" are examples of concepts for providing redundancy.

Technical redundancy is oftentimes realized through multiple systems assuming the same task. Depending on the significance of the task within the system, the changeover of systems will takes place more or less automatically.

Relating this to the human being, and particularly to the person in the cockpit, we are living today with a simplified form of redundancy. The simplest example of such a situation would be the complete discernable loss of a pilot. Yet, a functional redundancy in the event of "partial losses", overloads or errors will be dependent on many pre-conditions. A balanced working environment, an optimal hierarchical gradient, the ability to communicate effectively and Standard Operating Procedures (SOPs) must first be provided for, only after which is mutual monitoring possible. The key phrases here are Crew Coordination, Crew Resource Management and Communication.

3.7 Error Prevention in Practice

Accept your own susceptibility to error

Cognitively, our susceptibility to error is the reverse side of our most important attribute as pilots: our flexibility.

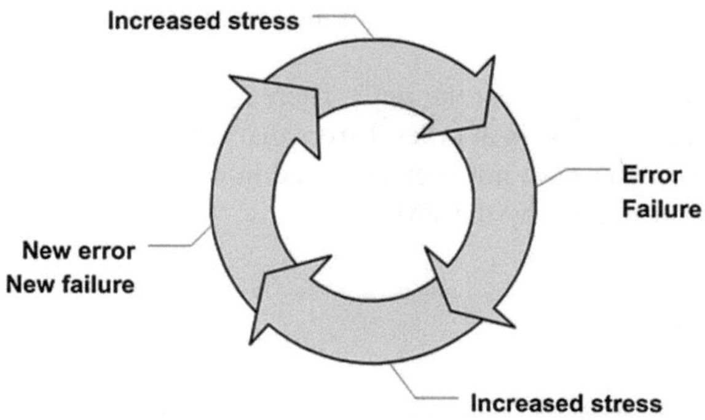

Fig. 3.8 Poor judgement chain

They are unavoidable in principle

Errors occur more frequently under time constraints, with a new task and when one overestimates his own capabilities ("young men make accidents"). Therefore, an attempt should be made to confront the emergence of error **preventatively** by:

- Avoiding time pressures
  Don't accept rushed approaches or an approach before completing its final preparations. If the aircraft is too high or too fast, a 360° manoeuvre should be flown, delay vectors requested or a missed approached executed.
- Theoretical knowledge
  All the "need-to-know" items should be readily available. The ignorance of procedures or limitations is unprofessional. Unknown situations, such as the first approach to a new airport, should be prepared for according to appropriate regulations.
- Dedication to the procedures

Deviations from SOPs are permitted only when they are absolutely necessary; and then only **after prior agreement.** A lax attitude towards the SOPs is not a sign of "expertise", but one of (potentially gross) negligence. All SOPs are the result of safety-relevant incidents; even accidents in most cases.

Poor judgement chain

If an error occurs despite the preventative measures, it will commonly be experienced as a personal failure. A failure inevitably leads to an increase in the level of stress. An increased level of stress leads to an increased susceptibility to error (see above), which quickly leads to the next error and raises the level of stress even further, which leads to yet another error. This vicious cycle is known as the "poor judgement chain" (see Fig. 3.8).

The increase in stress in this error chain can further lead to an unintentional disregard for SOPs and approach minima. Such has been observed in numerous incidents and accidents. For this reason, every unnecessary or non-agreed to deviation from the SOPs is a potential sign of entering into a dangerous error chain.

- For this reason, each *first* occurrence of an error must be addressed and rectified immediately.
- Never sweep an error "under the rug". There is no reason for this as long as the error prevention process is in order. Errors that then still occur are unavoidable.
- SOP deviations that have not been arranged in advance should immediately be called out immediately by the PM.

3.8 Summary

Everybody makes mistakes! This is the basic assertion behind our observations. In the course of evolution, the human has adapted himself optimally to his environment. In the process, he has not become error-free, but he has been able to survive superbly despite this shortcoming. Perhaps it is even a crucial part of his capacity for innovation because, as is well known, he is able to learn from his mistakes.

In the era of the Industrial Revolution, people began to take a systematic interest in their flaws. While the consequence of error prior to that time would have been limited to the erring person, himself, or, at the most, to a small circle of people, the use of machinery and technology meant that mistakes made by an individual would have ever greater consequences. Steam engines, railways and the beginning of the automotive society are a few examples.

This development has intensified considerably right into our own time. Just what consequences the failures of a few people can have on the entire human race was made clear through the accidents in Chernobyl, Bhopal[2] and Tenerife.[3] These accidents gave research into the field of human error a further boost.

We have discussed the classification of errors, whereby perhaps the most important breakdown into *active* and *latent failures* originated from James Reason.

Not only in aviation but in every industry, the spotlight focuses on that one person who formed the last link in the error chain. Of course, it is advantageous to have a perpetrator. Hardly a newspaper reader will be interested in reading about the complicated combination of circumstances. But, in order to still learn from mistakes today, it is important to not only spend time delving into causal research at the surface, but also to illuminate the background causes; an area where there has long been a significant deficit.

We have studied the potential for error associated with the cockpit working environment in a simplified model. The potential for error exists at every step along the path from the assimilation of information through its processing to the decision made as a result, right up to the action taken. It must be emphasized again and again that it is the human design, the way he functions, that allows him to make mistakes. At the same time, it is also this human design that makes it

[2] Poisonous gas discharges from a chemical factory in Bhopal, India in 1984 resulted in over 20,000 deaths.

[3] The worst accident in the history of aviation resulted in 583 deaths in 1977.

possible for him to be so versatile, intuitive and quick to react to unknown situations. One can't be had without the other.

We have shown that an error develops over many preliminary stages. It can be made while sitting at the desk, while developing an aircraft or in conjunction with the establishment of statutory rulings and regulations. It can come at the hand of company management, a department supervisor and, ultimately, from the mechanics, pilots or another person. In the context above, however, it is seldom just one individual error, but invariably an ominous combination of many such errors that lead to disaster.

It is therefore not sufficient to focus on just one position in the error chain in order to prevent that disaster. The overall system must be improved upon, and not just the last link in the chain.

The reduction of error and its related potential for disaster must be a continuous process. It is not possible to completely prevent latent failures, and while not flying at all still provides the highest degree of safety, it is not an option. Therefore, we must optimize the system.

Thus, both steps must be pursued:

- Errors can't be prevented altogether, but reduced.
- Although errors can't be prevented, their effects must be minimized.

Yet, with all the discussion about the human susceptibility to error, one thing should not be ignored:

Today, as in the past, it is the human being that, thanks to his unique abilities, is able to guarantee and increase the level of aviation safety in a manner that cannot be approximated by any machine. Any attempt to incapacitate, replace or curtail him will prove to be of only limited suitability. Development must not be directed against, but rather towards the person. It may be that the cause of 75 % of all aviation accidents can be traced in one way or another back to the pilots, but it is statistically impossible to determine just how many accidents were prevented by these very same pilots.

References

Bartlett FC (1932) Remembering: a study in experimental and social psychology. Cambridge University Press, Cambridge

Bohlman L (1979) Aircraft accidents and the theory of the situation, resource management on the flight deck. In: Proceedings of a NASA/industry workshop, NASA conference proceedings 2120

Eberstein (1990) Deutsche Lufthansa AG senior first officer seminar, Seeheim

Frese M, Altmann A (1988) The treatment of errors in learning and training. Department of Psychology, University of Munich

Green RG, Muir H, James M, Gradwell D, Green RL (1991) Human factors for pilots. Avebury Technical

Hawkins FH (1987) Human factors in flight. Ashgate, Aldershot

ICAO (1989) Human factors digest no. 1, ICAO circular 216-AN/131

Lufthansa (1999) Cockpit safety survey. Department of FRA CF, Frankfurt a.M.

Maslow H (1943) A theory of human motivation. Psychol Rev 50:370–396

Nagel C (1988) Human error in aviation operations. In: Human factors in aviation. Academic, New York
Nisbett R, Ross C (1980) Human inference: strategies and shortcomings of social judgement. Prentice Hall, Engelwood Cliffs
Reason J (1990) Human error. Cambridge University Press, Cambridge
Reason J (1991) Identifying the latent causes of aircraft accidents before and after the event. In: Lecture at the 22nd annual seminar "the international society of air safety investigators", Canberra
Roscoe SN (1968) Airborne displays for flight and navigation. Human factors
Tversky A, Kahnemann D (1973) Availability, a heuristic for judgement of frequency and probability. Cogn Psychol 5:207–232
Tversky A, Kahnemann D (1974) Judgement under uncertainty, heuristics and biases. Science 211:453–458
Wiener E (1993) Lecture at the flight safety foundation IASS, Kuala Lumpur
Williams J (1988) A data-based method for assessing and reducing human error to improve operational performance. In: Hagen W (ed) IEEE fourth conference on human factors and power plants, Institute of Electrical and Electronic Engineers, New York

Communication

4

Hans-Ulrich Raulf

4.1 Introduction

It hasn't been just since the development of the SHELL Model by Frank Hawkins (1987) that the correlation between cockpit crew performance and the interpersonal relationship of the individual crew members has been recognized for its significance.

Webster's New Collegiate Dictionary humorously referred to the cockpit as "a region noted for many conflicts" as early as 1961.

In his SHELL Model, Hawkins alludes to the interface between the acting person and his environment, thereby including the relationship to his colleagues as a possible source of error, as well. The foundation and origin of an interpersonal relationship is the communication between two people. Foushee (1984) also stresses the importance of the interpersonal relationship for the safe conduct of a flight: "Interpersonal phenomena can affect air transport operations".

The following discussion serves to demonstrate how effective communication decisively encourages both team spirit and teamwork. Communication is the link between highly specialized experts and guards against misunderstanding (Starck 1993).

The intent of this chapter is to describe a goal-oriented Communication training module as an independent component within the overall Human Factors training program. It also encompasses participant learning objectives, while the greater part of the chapter deals with the development and description of the training content. Recommendations as to possible training methods and follow-up activities are provided towards the end of the Chapter.

H.-U. Raulf (✉)
Main Airport Center, Vereinigung Cockpit e. V, Unterschweinstiege 10,
60549 Frankfurt, Germany
e-mail: uliraulf@t-online.de

H.-J. Ebermann and J. Scheiderer (eds.), *Human Factors on the Flight Deck*,
DOI: 10.1007/978-3-642-31733-0_4,

4.2 History

In the past, communication within the context of air transportation was considered to be merely "technical communication". Teaching content focused exclusively on correct phraseology, standardised commands for configuration changes and a simple "challenge and response" approach to the task-related use of checklists. At the same time, overall training focused on the individual.

The significance of crew coordination has gained prominence only since the 1990s with the introduction of "Line Orientated Flight Training" (LOFT) in the simulator.

At the same time, it is generally known that certain communication techniques form the basis for a safe working relationship: Standard Operating Procedures (SOPs) demand standardized communication practices. The "Sterile Cockpit" concept prohibits private conversations during safety–critical phases of flight.

The importance of interpersonal communication outside the bounds of the SOPs and during the critical phases of flight has been underrated. Yet, it is particularly by means of this "non-operational" communication that decisive prerequisites are created for successful teamwork in the event of an emergency or while under extreme time pressures.

Although this is addressed in today's training guidelines for cockpit personal under the rubric "CRM" (EU-OPS), the practical training content used to date is not sufficient for present-day requirements.

As a result of analyses of accidents with relatively favourable outcomes (UAL 811, Honolulu und UAL 232, Sioux City), Predmore (1991) concluded in recent years that exceptionally good communication between the crew members had a great deal of influence on the positive outcome of the respective situation. The study concluded that the working relationship within the crews (particularly in the case of UAL 232) was characterized by an efficient distribution of communication to the individual crew members in the form of so-called "task demands". In the process, each crew member was clearly assigned a specific task to accomplish.

Predmore identified a consistent pattern in the communication from the captain, who regularly alternated his focus of concentration to the different problem areas. To date, however, one crucial step has yet to be taken; namely, to precisely describe the essential characteristics that make up good communication while teaching the average pilot how to acquire and practice these skills.

This aspect will be pursued in the following discussion in the form of communication training, ideally oriented to the practical work experience. We do not claim to be presenting anything fundamentally new; much more, we have taken building blocks from various publications and have combined them into a comprehensive training concept.

4.3 Tasks and Targets of Communication Training

> In groups with highly structured tasks and powerful leader positions, such as the cockpit crew, task-oriented leaders performed better as long as interpersonal relationships within the crew remained relatively good (Foushee 1984).

Communication training within an integrated Human Factors training program has the task of creating the interpersonal basis for team-oriented behaviour. The targeted application of personal behavioural strengths or the avoidance of unfavourable behaviour patterns, as the case may be, is fostered.

In addition, communication training should facilitate an awareness of what the individual can do to enhance the optimal functioning of a team.

According to Hackman (1993), the so-called "low workload phases" of a flight should be purposefully used to integrate the individual member into the overall team.

Particular emphasis during communication training is also placed on team development: How important is the first impression and what can an individual do personally from the outset to help get the team off to a good start? How does a person create a good working environment? Specific behavioural analyses and tips are needed to assist the individual in such endeavour.

Hackman claims that it is useless to attempt to train all pilots toward a specific behaviour pattern. Under the influence of excessive stress, each person will fall back into the behaviour patterns they acquired early in life and will no longer be capable of displaying the behaviour pattern recommended during the theoretical instruction.

From this it can be concluded that, following a high-stress situation, only conscious effort can help bring one's own behaviour back into an effective working mode. This leads to an awareness of one's own style of communication.

In preparation for such inherent stress situations, one can deliberately and consciously revert back to acquired capabilities (so-called skills), meaning that the personal style of communication with other crew members (or engineers, passengers, ATC controllers, etc.) should be varied to adapt to the situation.

An awareness of certain communication styles can be subsequently linked to potential conflict situations. These conflicts can be taken on and resolved using appropriate strategies and skills. This further develops the interpersonal capabilities of the individual.

The process of self-evaluation during training works not only to highlight the weak areas associated with one's own behaviour, but it also exposes many positive behaviour patterns already present in many of the participants that have not been deliberately applied. The objective is to bring out and reinforce these strengths through explicit feedback.

The recognition of one's position within the occupational hierarchy is the starting point for a working relationship. An atmosphere of mutual respect and trust begins with an appreciation for the colleague's capabilities and with personal estimation afforded on an interpersonal level. The often cited "good working atmosphere" establishes the foundation for the safe and effective functioning of a crew.

4.4 Verbal Communication

4.4.1 Truthfulness and Effect Awareness

As an introduction prior to actually getting into the theoretical content or practical exercises, we consider it sensible to first touch upon the balance between "truthfulness" and "effect awareness". This can be done very impressively with the help of the Square of Values according to Schulz von Thun (1989), which is developed exemplarily together with the participants (see Fig. 4.1):

Truthfulness and authenticity deteriorate into naive bluntness if they are not paired with an awareness of the effect. This can either be indicative of a lack of tact or implies a very little appreciation for tactical necessity. The polar opposite to this is the "overall diplomat", whose every word is cast as tempting bait and whose calculating rhetoric has become second nature: the true essence remains hidden; not only to the dialogue partner but to the speaker himself (Schulz von Thun 1981). From this, the crossover to the individual objectives of each participant also emerges: The progression from the Square of Values to a Square of Development (see Fig. 4.2) should identify stimuli that point in the direction where development makes sense and do justice to the special challenges faced by the particular professional practice.

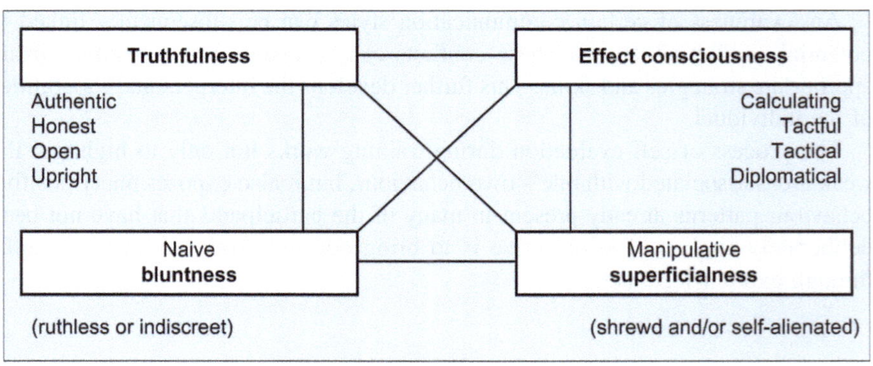

Fig. 4.1 Square of values

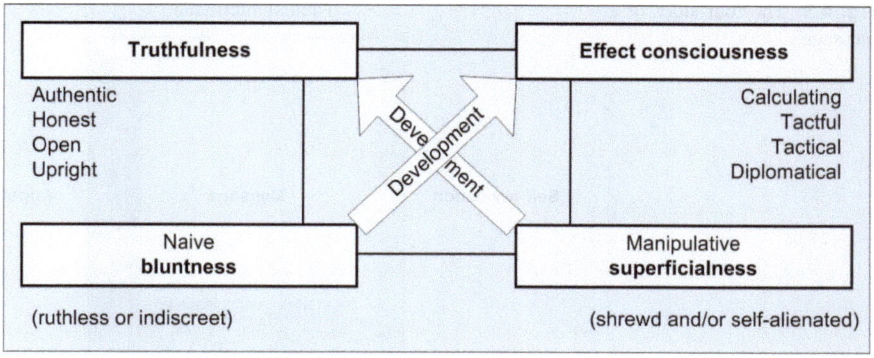

Fig. 4.2 Square of development

4.4.2 The Message Content

Schulz von Thun (1981) distinguishes between the four different aspects of a message that are simultaneously in play and "emotionally operative" at all times.

- **Factual content** contains information to be conveyed about the objects and processes of the world.
- **Self-revelation** contains information conveyed by the "sender" about himself.
- The **relationship indicator** discloses how the sender views his position with relation to the receiver.
- The **appeal** is an attempt to exercise influence in a particular direction.

In this context, it is of further significance that the elements of this message are not only consciously or unconsciously "packed" inside the sender's words (and gestures), but they also influence the interpretation by the receiver. Because of this, it is entirely possible (and normal) that one and the same phrase can be taken in and assessed differently by two different dialogue partners.

Schulz von Thun (1981) depicts this in his Square of Communication (see Fig. 4.3), which is a further development of the model from Bühler and Watzlawick.

In an explanation of his model, Schulz von Thun emphasizes that the sides of the square all have the same length: the four aspects can be seen as equal in ranking, even when one or the other aspect stands in the forefront of any particular situation. An over-emphasis on the factual aspect in professional life, however, stands opposed to this notion. In the opinion of many pilots, only factual content counts in the cockpit, thus relegating cockpit communication to a value-free and unemotional exercise. This suppression of the other aspects of a message often leads to one or more relationship conflicts being carried out at the supposed value-free factual level, thereby compromising the quality of the working relationship.

It is essential to convey to the participants that certain inner conditions and personality biases are connected to the preferability of particular aspects of a message and correspondingly shape the style of communication.

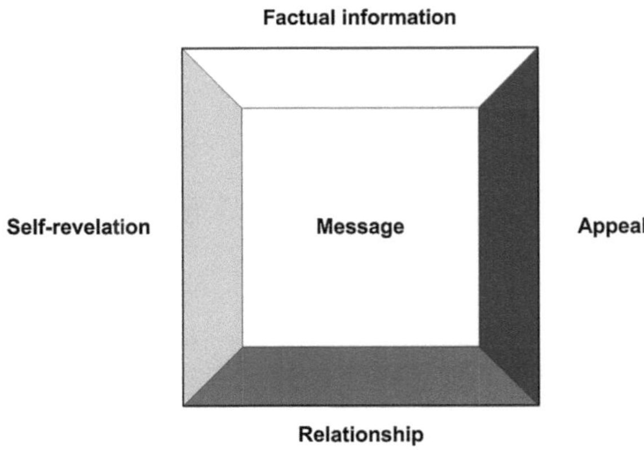

Fig. 4.3 The four sides of a message

Factual content

In most cases, factual information stands in the foreground of the message. It can relate, for example, to perceptions or organisational processes. "Standard callouts" in the cockpit are a good example of this: "Speed is low" or "One dot high", etc., are open references to specific deviations from target values.

Self-revelation

Additionally, every message contains information about the personality of the sender, as well. He uses various techniques (e.g. self-exaltation or self-concealment) in order to "portray himself in the right light". This could mean that a standard callout contains the unspoken cue, such as: "I am alert as PNF and am actively participating in the conduct of the flight".

Relationship indicator

Just where the sender stands with regard to the receiver—what he thinks about him—can be further drawn from the message. Sending a message always means that the type of relationship is being expressed to the person addressed, as well. This can oftentimes be seen in the choice of wording, the inflection of the voice and the non-verbal secondary signals. The receiver lends a particularly sensitive ear to this part of the message, because this is where he feels treated or "mistreated" in a certain manner as a person. The counterpart is directly implicated—"affected"–by relationship indicators, and oftentimes in a manner other than intended.

This aspect of communication is comprised of two types of messages: For one, there is the sender's estimation of the receiver ("How do I see you?"), and then there is a definition of the relationship ("This is how we interrelate to each other!").

Using the example of standard callouts, the captain, as PM, may wish to express his positive intentions, such as: "I want to help you; we are a team and can work together!" or, with degrading intentions: "Your flying is not good enough; you'll never be able to get down without me!".

This can lead to a disruption in the teamwork if the receiver attempts to defend himself against the relationship indicator. He won't reject the factual message, but may protest at the relational level: "That's true, but what difference does it make?"—he feels "harassed".

Appeal

There is very little said just "incidentally"—almost all messages have the function of trying to influence the receiver. Thus, the standard callout should trigger the correction of a deviation. An attempt to exercise influence can be more or less done in an open or concealed manner (manipulation).

The manipulative sender does not shy away from also calling into service the other three sides of the message in order to influence the outcome of an appeal. Reporting from the factual side is then one-sided and tendentious. The aim of self-projection is to affect a certain impression on the receiver, while relationship messages are defined by the objective of "keeping the other's spirits high". If the factual content, self-revelation and relationship indicators are aimed at improving the outcome of an appeal, then they will be functionalized, meaning they do not reflect what is, rather they become a means of achieving what is intended.

4.4.3 Congruency

A message is congruent when all signals point in the same direction, meaning the overall message is consistent in itself. If the verbal and non-verbal signals do not match, however, then it is unclear to the receiver, in which of the two aspects the actual intent of the message is contained; such communication is referred to as a "mixed message". The receiver automatically values the non-verbal signals at a higher level than the spoken word: "He did in fact mention that..., but I don't know why I didn't believe him!"

4.4.4 Communication Styles

Although Schulz von Thun (1981) differentiates between a total of eight different styles, we consider it appropriate for our purposes to reduce that number down to only four styles (see Fig. 4.4). This was developed by Sherod Miller at the University of Minnesota and translated for use in the German language by Paula (1992).

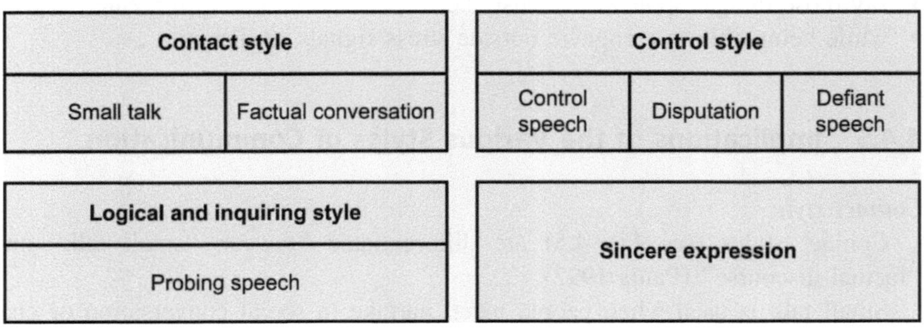

Fig. 4.4 Communication styles

A more detailed description of the individual styles can be found in the section "Effects of the various communication styles" below.

The message contained in each statement is comprised of two parts: the factual content and the style in which it was conveyed. Analogous to Schulz von Thun's Square of Communication, factual content can also contain various aspects.

Factual content can relate to a broad variety of topics: overtime, sports, mechanical failure, plans, weather, private matters, etc. The other person is notified as to how the message should be received based on the style used. Style encompasses both the non-verbal body language and the verbal aspects of a message, while expressing subliminal intentions, as well.

Each style has its own characteristic intentions, activities and forms of speech. The message recipient becomes attuned to the intentions of the sender through the style used. Paula makes it clear that the content of a message is "assimilated" through the mind. The "How" is discerned through the eyes and ears.

The recipient very often interprets the message based on the style of communication. What one person intends as control speech may for all practical purposes be interpreted by the recipient as disputational speech. "The meaning of a message lies in its impact, not in its intention" (Paula 1992).

Capt. Gunnar Fahlgren (1989) proposed the theory that 60 % of a message is transmitted non-verbally, 30 % through the style of communication and only about 10 % through verbal expression. The style of communication is controlled by subconscious mechanisms that cannot be disregarded. Therefore, it is important to be aware of one's own communication style and to vary it as needed to adapt to the respective situation. Thus, every style will be used and there is no one style that is always correct or that can be used at all times.

> He who knows communication styles has taken the first step towards better understanding (Paula 1992).

This awareness should enable the participant to

- communicate more flexibly and
- say those things, upon which trust and respect are based, with greater ease
- while being able to recognize outside stress signals earlier on.

4.4.5 Implications of the Various Styles of Communication

Contact style

Contact styles (see Fig. 4.5) are differentiated between "small talk" and "factual discourse" (Paula 1992).

Small talk is used when people meet, partake in social conversation or chat prior to more serious conversations (e.g. party-talk). Factual discourse serves the

Fig. 4.5 Contact styles

Contact style	
Small talk	Factual conversation
Friendly Sociable Relaxed Entertaining Contact building	Information gathering Information sharing Workflow oversight Networking

routine exchange of information (e.g. flight relevant information). Both forms are used quite often in the day-to-day interaction between people.

Small talk

This friendly, conventional and oftentimes playful style is the starting point for most conversations. It is useful for getting mutually acquainted (e.g. with the greeting) as well as for initiating more intensive discussions. Circumventing the elements of small talk when just getting acquainted and moving directly to another style is commonly deemed as "not being according to convention" and may lead to uncertainty (e.g. "charging the gate like a bull").

The dialogue partner uses small talk to be friendly while probing the state and the nature of the other person, as well as to avoid "making waves". These types of conversations deal mostly with the weather, news, sports and other common areas of interest. This is where a bridge is built between the dialogue partners that can later be used during conflict situations.

Effect

Small talk is only a waste of time when this preparatory level of conversation is not put aside at some point by the dialogue partners and impedes entry into the factual conversation. Chatting conveys information about oneself, establishes contact with the partner and builds trust. What's more, small talk is well suited for the pursuit of the following objectives:

- One can alter the pace of work, relax a bit and find rest in the presence of the other person (e.g. "celebrate" one's own achievements, sit back with satisfaction).
- One receives clues as to the other's mental, physical and emotional state. One learns to what extent he can get to know the other person and whether it is the right moment to address a certain topic.
- Small talk is often used as a "warm-up" at the beginning of a mutual flight schedule.
- One can reduce tension and avoid or test certain technical matters, allowing sensitive issues and potential conflicts to be carefully explored.
- The non-verbal behaviour of the dialogue partner can be observed during an informal conversation, while the interpersonal mood and atmosphere can be discerned.

Small talk is not suitable for serious technical questions or for resolving conflicts.

Nevertheless, small talk should be actively implemented in order to build a relational platform as quickly as possible.

Just how important this relational platform is was affirmed by a NASA study, in which it was demonstrated that crews who had already flown together for several days made considerably fewer mistakes in complex simulator scenarios than crews who had just become acquainted with one another. This, despite the fact that they were tired and the "new" crews were well rested (Foushee et al. 1986).

Factual discourse

This is the style used to take care of daily business. Information required for completing a task is gathered and disseminated on a routine basis. Typical attributes would include reports, observations, facts, enquiries, checks and even minor decisions.

Factual discourses are initiated through open questioning, meaning that an answer of simply yes or no is not possible (who, what, where, when and how).

> This is the prevalent style of communication when working in the cockpit, especially during periods of low workload.

Effect

Regular routine communication promotes contact with colleagues and generates information regarding their personal commitment, attitudes and opinions.

The one who keeps his eyes and ears open at all four levels of communication will be in a position to recognize technical and interpersonal problems early on, and to take corrective action accordingly. The message conveyed by a good supervisor is: "I am well informed and will look after the concerns of my colleagues".

Control Style

This style deals with control and persuasion; in other words: Power. This can be achieved through three different methods (see Fig. 4.6):

Control style		
Control speech	Disputation	Defiant speech
Steer Initiate Appraise Set expectations Attain agreement Reinforce the positive	Force changes Defend one's self Conceal Intimidate Bluff	Protect one's self Fight back Convey guilt feelings Seek revenge Hide offense

Fig. 4.6 Control style

- the control discourse
- the disputational discourse
- the defiant discourse

While the intent of control speech is altogether positive and constructive, both of the other methods can have negative and destructive intentions and effects. Yet, all three types of this communication style have so much in common that they can be easily confused; displaying the following exemplary stylistic devices:

- should, must (indefinite)
- leading questions
- always, never
- you statements
- why questions
- command forms

All three methods have the same intent: to control and to persuade. They commonly appear in far more subtle forms and are presented somewhat exaggerated here for the purpose of illustration.

The difference lies in the intent:

In control speech, the other person is incorporated in a productive and constructive manner. Both of the other types are offensive and destructive.

If a consensus cannot be achieved through control speech, then a participant's discourse can easily escalate to disputational speech; the conversation gets louder and harder, and the pace quickens. Its defence deteriorates into defiant speech, where the only things that matter are victory at all costs and the protection of one's own self.

> Therefore, it is very important to acquire a gentle touch when it comes to the use of control speech in the cockpit and to be able to immediately recognize any drifting of the discourse into disputational or defiant speech.

Control speech

This style encompasses the issuance of instructions and the stimulation of action. It is the style primarily used by the captain in the cockpit (or the pilot flying, as well, depending on task distribution). It is characterized by activity and effectiveness, not by a passive waiting and reacting. Good instructions are short, precise and effective. If more information is required by a colleague, however, then it is better to pursue this in a logical and inquiring style. Using this method:

- Activity is steered
- Borders and limitations are removed
- Progress is measured
- Instructions are issued
- Desired behaviour is reinforced.

When correctly applied, this style develops respect, trust and authority without the need for the speaker to insist upon compliance.

> Control speech is well suited to the leadership of a team.

When colleagues are included, consensus and cooperation will be achieved. Otherwise, antagonism and rejection may arise.

This style has achieved its purpose when the instructions are carried out and the established goals are consequently pursued. Approval and consensus are most readily encouraged when instructions are appropriate to the situation and not imposed.

The danger with control speech is that pressure develops rapidly when consent is lacking and both dialogue partners can fall easily into a disputational situation. The control discourse should be interrupted at this point and a switch should be made to another, more productive style. The transition to another style is generally very important because the exclusive use of control speech tends to neglect the human side of the working relationship. The speaker is perceived as being a persistent, bossy "drill sergeant", from which rejection and defiance gradually emerge.

Effect

As implied in the previous section, control speech is particularly dependent on "How?" it is used. The productivity and contentment of all parties can be realized only when both the individual's own self-esteem and his respect for the colleagues are preserved.

Almost everyone welcomes and accepts instruction when it is needed. Only those people with anti-authoritarian personalities, as described in Chap. 6 under "Disturbance factors", may have problems with it.

If the goals are unclear, or where the working relationship is disorganized or uncoordinated, restlessness and uncertainty can develop, but this can be resolved through the issuance of appropriate instructions. If a colleague is degraded due to disrespect, however, it could be indicative of inappropriate dominant behaviour—whereupon control speech would no longer be used but disputational speech.

Disputational Speech

Characteristic for this type of discourse is the direct, aggressive and oftentimes degrading language.

The speaker doesn't place himself at the level of his dialogue partner, but rather distances himself from every commonality in an attempt to injure that person. It is no longer possible to proceed at a factual level; emotions gain the upper hand. The person desires to impose change by forcing the other to abandon his own standpoint. In the process, the discussion slips into an uncontrollable "ping-pong" match, without hope of making progress or reaching a consensus.

In response to (frequently "below the belt") accusations, one merely defends his personal standpoint without having the chance to effectively develop it. The speaker is no longer open and amenable to the argument from the counterpart who, at this moment, is not a dialogue *partner* but an *opponent*. His anger is aimed at hiding his own fears. Blame for a project failure or for the conflict itself is put off on others. The greater the fear, the lower the degree of trust and effective communication will be. Even the simple transfer of information will miscarriage.

Trying to force a solution, issuing orders, passing blame, threats, attacks, provocations and insults represent typical situations conducive to disputational speech.

Paula (1992) notes in this regard that: "Even the most justifiable assignment of blame is senseless". When a person feels his self-esteem (skills, judgement, capabilities) being attacked, it is very probable that he will then attempt to reimpose his superiority through disputational speech.

Effect

Disputational speech is a sign of stress for all involved: the speaker has lost control over himself! Everyone involved concentrates solely on the others and their weaknesses, while disregarding the weaknesses of their own. From this an ever-more-rapidly spinning merry-go-round of accusations evolves. The result will be apprehension and rejection in most cases. The work within the team is enduringly and significantly disturbed over the long-term.

> When a slide into this style becomes apparent with regard to work in the cockpit environment, it is absolutely essential that it is recognized immediately by all parties and that the basis for a new start is pursued without delay.

This is made all the more difficult because the disputer will have only limited access to "reasonable" arguments.

Defiant Speech

> If disputational speech is a war with words, then defiant speech is guerrilla warfare (Paula 1992).

In comparison to the openly carried out disputational dialogue, defiant dialogues are indirect and covert. They are an attempt to secure a position of power, starting from a position of powerlessness. In the process, the control role is exercised "from the bottom up". The person shuns responsibility and, in so doing, undermines any effort on the part of others to implement change. An attempt will also be made to passively convey a sense of guilt onto the others.

Signs of this style are detected in deep sighs (of frustration), a slouching posture, a lack of strength, non-committal responses, changes of subject and silence. The speaker degrades himself and his capabilities, consequently leaving himself incapable of affecting a solution or a change.

Defiant speech is an indication of the often discussed "inner resignation".

> In the cockpit, it is an unmistakable indication that the foundation for effective and safe teamwork has crumbled.

Effect

Defiant speech wastes energy and information; productive work has been largely sabotaged. The speaker damages himself through continuous degrading. Cooperation has become impossible.

Logical and Enquiring Style

Brainstorming—has become a buzzword in the English language and provides access to this style (see Fig. 4.7).

Problems arise during routine work that can't always be resolved by the simple acquisition of information during a technical discussion. Background information must be checked, technical questions answered and alternatives developed.

All parties will be challenged and encouraged to analyse and interpret the information, as well as to submit recommendations. Opinions will be called for but not assessed. This style is a successful coupling of the positive aspects found in control speech with the "sincere expression" described in the section that follows.

An important prerequisite for properly using the logical and enquiring style is the capability of being a good listener and observer. Non-verbal information in this context can provide important clues.

> This style is well suited to the preparation needed for making a good decision in a difficult situation.

But beware: if the time available is limited, not everyone will have a chance to respond. The captain must then take the initiative and establish the "course of action" according to adequate information acquisition. Further information regarding the assessment of action alternatives can be found in Chap. 6.

Effect

When using this style of communication, everyone will get the feeling that they are being taken seriously and treated equally.

One level still remains to be factored in, however: the emotions. What the participants' actual motives are, what feelings are motivating them and how they envision their own desires are not addressed. Problems at this level, and especially the potential interpersonal friction within a team, remain unresolved. This can lead to personal dissatisfaction.

It becomes particularly dangerous when everyone stands firm at a non-committal level with nobody willing to make a commitment or accept

Fig. 4.7 Logical and enquiring style

Logical and enquiring style
Maintain overview Determine cause Create alternatives Assess options Seek advice

Fig. 4.8 Sincere expression

Sincere expression

Being responsible
Bringing together, not controlling
Respecting one's self and others
Working together
Caring, sympathizing
Understanding

responsibility, meaning an absence of tangible results. Verbal signs of this are expressions such as "one could perhaps..." or "another possibility might be...". This is usually indicative of difficulties existing at other levels, yet with everyone avoiding the attempt to resolve the core problems.

Sincere Expression

When it becomes obvious through the logical and enquiring style that the attempt to solve a technical problem is failing due to tensions or differences of opinion, then it is time to switch styles (see Fig. 4.8).

In addition to external information, internal information is also queried: What emotions were not factored in at the logical and enquiring style or were allowed to gain control over one's own actions, undetected, at the disputational or defiant speech levels.

Sincere expression neither assigns guilt, nor does it build a defence or make demands. The cards are dealt face up on the table, without tricks. An attempt is made to build a constructive working relationship or to protect one from being destroyed.

Typical situations calling for sincere expression are those related to interpersonal conflicts and negative feedback (=criticism), as well as the elimination of tensions and barriers when solving a problem.

Characteristic for this are commonly the "I" phrases, as opposed to "you" phrases, with respect to one's own perceptions or emotions (e.g. "I have the impression that ..."/"I feel that I am not being taken seriously ...").

Paula identifies prerequisites for the correct usage of sincere expression as being:

- the recognition of one's own inner and outer worlds
- the acceptance of existing circumstances
- the actions based on one's own consciousness

Effect

Sincere communication can be firm yet considerate, determined yet flexible, caring yet controlling; all at the same time. Everyone is accepted and appreciated as a human being. The core of a problem is identified during the dialogue while desires and emotions are addressed at a level deeper than merely superficial. Tensions and differences of opinion are addressed openly and solutions acceptable to all are pursued jointly. Participants clearly communicate their preparedness to partake in open discussions and do not insist on the absolute assertion of their own will. The most important characteristics in this context are that, not only are remedy *Options* being sought after, but responsibility is assumed and commitments are made. The participants go on their way with practical agreements.

In this manner, trust can be created and a good basis for safe and effective collaboration (re)built. Everyone on the team can do their part to contribute to it without fear of being degraded, injured or penalized for their comments. No one has to hide, defend or protect himself.

Nevertheless, there are limitations to the use of this style. Where trust is not present, sincerity is out of place. The protection against misuse of the information disclosed becomes more important at this point. This is the cross reference to the "Truthfulness and effect awareness" in the Square of Values developed at the outset. Abuse in the form of a targeted exploitation or the pretence of sincerity in order to instrumentalise behaviour should also be rejected: Whoever would use sincerity for the purpose of manipulation will quickly experience a "belly landing" and create even greater distrust.

4.5 Non-Verbal Communication

> We spend time and energy to learn other languages in addition to our mother tongue. Over time, body language has become recognized as a foreign language, as well. We do not necessarily have to learn foreign languages, but we get a lot further ahead if we do have a command of them. We reduce the risk of a misunderstanding (Molcho 1983).

This section is prefaced with the remark: The following is not an attempt to take psychological laymen and transform them into psychotherapists; able to analyse their counterparts on the basis of movement or expression.

It is merely intended to portray the great effectiveness of non-verbal communication. For, when differences do exist between verbal and non-verbal information, the receiver automatically places higher value on the non-verbal information. To this end, a process of "self-observation" should be set into motion.

Which of my behaviour patterns are favourable and which are unfavourable with regard to teamwork?

In addition, it should be recognized that body language is above all an expression of our perceptions (Molcho 1988).

An important point to address at this point is the "first impression": Every person is more or less inclined to unconsciously make a "picture" of the other person.

Many of the impressions from the first minutes of an introduction reinforce this picture so strongly that it may be impossible to correct it later. "Emotional information is more rapidly implemented than linear information, is quicker to trigger a reaction and opens a wider spectrum of interpretation" (Molcho 1988).

The advice, *not to pass judgement based on the first impression*, may be well intentioned, but it is only of little help. We can't always divest ourselves of this influence because it largely plays a role in the subconscious mind.

The perspective should not be directed toward the counterpart, but toward one's own mirror. The "lasting impression" will be made during the first introduction, which, in the flying business, almost always comes with the initial briefing. The captain sets the tone and the priorities for the subsequent working relationship—and that, without saying a word. If he pulls the flight plan towards himself or positions himself in a way his colleague can't read it, then he has (perhaps) already sent out the first signal: "Your opinion is not called for. I, alone, am solely capable of evaluating it!"

His colleagues will then most likely refrain from sharing their knowledge and their experience, even when he verbally invites their comments.

Non-verbal communication reveals itself as a domain in which one thing in particular applies. It can't be learned in a seminar; it needs to be experienced personally: one's own style, one's own movements. Photographs depicting exaggerated examples can be helpful with this. Samy Molchos' books are alive with these exaggerations, portraying striking examples.

For an estimation of one's own personality, it is most effective for the person to view himself in the mirror or with a video (LOFT).

4.6 Technical Communication

4.6.1 Radio Telephony

Many accident and incident reports confirm that radio communication, notorious for its interference problems, represents a significant source of misunderstanding.

While there may be sufficient capacity during normal operations for detecting error, this changes drastically under stress. The frequency is jammed; there is little time for instruction and readback; the workload is increasing. There is only one approach that promises success: strict "discipline".

This begins in the quiet, stress-free phase of flight preparation. From start-up request through engine shutdown, the consistent use of standard phraseology, alone, will warrant the successful outcome of the flight. Only what is consistently put to use can become second nature and will stand up under stress, because it no longer requires specific mental activity: "How do you say that correctly?"

It may sound very professional when a pilot bids farewell to the approach controller with "XY six-forty, thanks, good day!", but is he actually descending down to 5,000 ft or possibly down to 4,000 ft? Is he really flying the approach to Runway 25L or perhaps to 25R?

The consequent readback of clearances not only facilitates the monitoring by colleagues in the cockpit and the controllers, ensuring that what was said is also that which was understood, but it also enhances the situational awareness of the crews in other aircraft and enables them to "plot a bigger picture" of what is going on around them in the sky.

4.6.2 Marine Concept

The same concept that works for radio telephony (R/T), applies here as well: Procedure-related communication is set forth by the manufacturer and the aircraft operator. It should be binding, clear and unambiguous for all parties. A "coordination concept" and related phraseology can be developed unhurriedly at the desk; removed from any hectic distractions. The day-to-day experience gained in working with this concept should be fed back into a process of on-going development, thereby representing a form of validation.

In practice, such a concept is particularly helpful during those phases of high workload: "Please lower the Dunlops" may be humorous, but certainly not professional: "Gear down" is short, clear and unmistakable—if it is used in a disciplined manner.

The creation of a sensible concept for standard communication is especially important for acceptance among the crew members. This concept represents the first link in the error chain (see the section on "Human error and reliability").

If the concept is thrown back and forth time and again (e.g. checklist answers *Checked* instead of *Tested* and vice versa), then it will not be accepted—no matter how disciplined!

4.6.3 One- and Two-Way Communication

One-way communication in this sense means communication in only one direction. The speaker receives neither response nor sign from his colleagues that they have heard or understood anything.

For this reason, two-way communication is always preferred. It constitutes a form of interpersonal communication parallel to feedback. I confirm what I have received and now give my colleague the opportunity to correct any misunderstanding: Readback.

The "acknowledgement", as well, also fulfils an important task. The recipient sends the signal: "Yes, I am following you. I am involved and know our position with respect to the progress of a procedure or a phase of flight". This prevents uncertainty on the part of the speaker and frees up his mental capacity for the next task.

Both the "Readback" and the "Acknowledgement" must be integrated into their mandatory procedures by the airline. This is already the case with radio communication, and the fictive "XY Six-forty" crew above would be well advised to also put it into practice.

Two-way communication serves another important function by facilitating the detection of incapacitation.

4.7 Contractual Role and Personal Estimation

To begin with, the division of tasks in the cockpit is very clearly regulated. The captain is the supervisor of his crew and bears ultimate responsibility. The members of the crew work at their assigned duties, either independently or with two-way monitoring.

Yet as self-evident as this might be, it can also lead to problems. This is often the case when the assignment of professional duties (contractual role) is accompanied by a personal estimation, meaning that a form of "human hierarchy" is derived from the professional hierarchy.

Both the supervisor and the person being supervised are susceptible to this false conclusion. The boss who envisions himself as being surrounded by incapable and unwilling "individuals" is quick to seize the control style of communication when giving instructions.

Because of frustrations that his colleague may not be capable of achieving the same quality of work, his personally degrading remarks may also be mixed in with the dialogue. The colleague has a very sensitive ear to this message and feels further intimidated by this "man of action". The work of the entire crew is then jeopardized by this breakdown at the interpersonal level.

Problems of this nature can also originate with the person being supervised. This may be the case if he feels his own capabilities are lacking and believes he can't accomplish anything without the supervisor. If he degrades himself from the outset, then he, as well, creates a dangerous slope, or gradient, within the cockpit.

This may play out to the other extreme, as well, when an FO overestimates his own capabilities; deeming the captain to be incapable and being forced to assume the role of "stealth captain". This, too, establishes a gradient within the cockpit, against which the captain must invest all his energies to renew the proper relationship.

In either case, both the supervisor and the person supervised must use an enormous amount of their mental capacities to rectify their personal position within the role structure, which is then missing from the concentration available for the task at hand.

This forms the basis for the "Trans-Cockpit Authority Gradient" as a slight, yet ever-present slope between the crew members (see Fig. 4.9), but should be differentiated from the relationships at the interpersonal level between crew members.

This is where mutual respect and appreciation are necessary that don't depend on the capabilities of the individual. All players are at one level. Everyone is appreciated as a person while every opinion is heard and considered. This establishes the prerequisite for "assertiveness"; the capacity of a colleague to freely and openly express his opinion in critical situations or to forcibly defend it when safety is at risk, as the case may be.

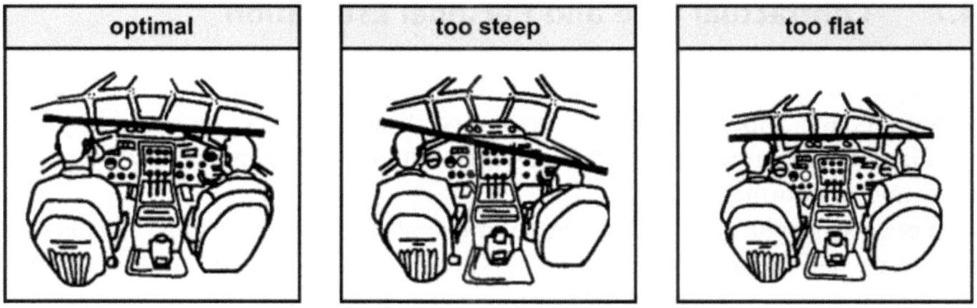

Fig. 4.9 Trans-cockpit authority gradient (according to Hawkins 1987)

4.7.1 Strategies for Conflict Resolution

No matter how well intentioned the efforts are for being a good team member, it won't always be possible to avoid friction and misunderstanding.

As discussed in Chap. 3, it is essential to not only avoid interpersonal conflict and misunderstanding, but to be able to function despite these disturbances, as well. Merely recognizing that something is wrong in the working relationship is not enough. The attitude commonly encountered that "it's just not worth quarrelling with this colleague because we'll go our separate ways after 2 or 3 days and (perhaps in a large fleet) may not see each other again for years anyway", must be corrected at the source.

Each conflict not only affords the risk of getting angry or being injured, but also presents an opportunity to grow through constructive debate. Furthermore, in the interest of a safe working relationship, it is essential to continuously strive to establish a starting point for good cooperation: The next stress situation won't wait until we have the opportunity to fly with a likable colleague once again! Moreover, as was mentioned at the beginning of the chapter, a good working atmosphere and mutual trust make up the best prerequisites for successful action in exceptional situations. It is necessary to develop a strategy and put together a type of personal checklist which can be called upon to approach difficult situations with. This strategy relates to the relationship and role conflicts that have arisen due to certain repeated patterns of behaviour. It functions only during phases of low workload and is not suited for use when stress or time constraints exist. As such, it is intended to be preparatory for good teamwork under stress.

This strategy comprises seven steps (Paula 1992):

Recognition and analysis of the subject

Many individual units of information or messages, when put together, produce the situation—but what is the real "issue"? Where does the actual problem lie? What messages have been conveyed at the factual level; at the relational level? What have the participants revealed about themselves; what is their appeal? Putting all this information together then makes it possible to arrive at the real issue. At the same time, it is important to be aware that all the information taken in is subjective and that we, too, are still selective. What doesn't fit into our picture will be filtered out unconsciously. The core of a matter is more closely approached in mutual consultation than through personal presumptions.

Establish procedural rules—consultation

This has to do with the organisation of a discussion: when, at what location, who is to be involved, how long? The clarification of procedural rules by mutual agreement is oftentimes already the gateway to understanding. The subject can be changed and it becomes obvious that it is still possible to come to an agreement with the counterpart—a new level has been reached.

Discuss the matter

Only afterwards is the actual problem then discussed. One's own point of view is presented along with the information gathered. Everyone has the opportunity to point out their concerns. At the same time, it is important to listen attentively. Unclear or ambiguous information should be rechecked. Enquiry should be made just as with R/T: "Say again—please repeat, I did not understand that!"

Interpretation will not be made and a guilty party will not be sought after, but rather a common basis and mutual understanding will be pursued.

Clarify intentions

What am I concerned about? What do I want to achieve? What are my desires—for myself, for the team, for the situation? The more clearly each person regards these questions, the more clearly he will be able to express himself to, and make himself understood by his colleagues. If, on the other hand, this same information from the dialogue partner is not available, then specific related questions should be asked—and not left up to guesswork!

Develop alternative courses of action and determine the next steps

It is appropriate at this point to recall the differences between the logical and enquiring style and sincere expression. This dialogue should not come to an end without mutual agreements and commitments—everyone contributes his part to the overall solution.

Check the results

"Out of sight—out of mind!" This phrase should not apply here! The results of the discussion should be subsequently checked, as well: Have I kept my word regarding my commitments? Has the situation improved? If not, what is the real issue? How is the working relationship since the discussion? Have we gotten stuck in the same old rut once again; fallen back into the same behaviour patterns? These questions should be clarified jointly.

Metacommunication

The discussion about the discussion: Have we reached a different, better level of cooperation? Do I now understand my colleagues to a greater degree? How have I been doing since this discussion?

4.7.2 Feedback

Feedback is the most important tool available for an early recognition of, and perhaps even the prevention of misunderstanding, while enhancing the partner's capacity to understand at the same time. In commercial aviation, the "briefing" has become institutionalized over many years, while the "debriefing", in contrast, has been misunderstood to be an occasion for a social get-together at best. It is essential that this practice be permanently anchored in the crew consciousness so that every flight mission will be discussed. In so doing, it is important that an appraisal be made not only "from the top down", but that everyone is allowed to input their views and experiences.

4.8 Communication Proficiency

The points of friction in the SHELL Model, or the interfaces between the individual sections, are understood to be potential sources of misunderstanding.

Yet, what can be done to make communication as safe, as unambiguous and as clear as possible in order to reduce misunderstanding?

The technical environment, or the situation in the cockpit, should be considered at first: noise, stress, time pressure and an enormous information density make it difficult to appropriately express the communication skills developed in this context. Nevertheless, there are options that can be put to relatively simple use:

4.8.1 Express Yourself Clearly

Loud and clear—actually self-evident given the high level of noise in the cockpit; yet, with the (boom) mic right up to the lips, it may seem unnecessary at times to speak loudly—technology will take care of it. But when something needs to be conveyed to a colleague, it may be easily forgotten that the supporting technology is not available at that moment.

Yet, attempts can be made towards improvement at the content level, as well: What do I really want to say? What do I want to achieve? Should I first try to clarify these things when opportunity allows. It is essential that mixed messages are avoided and congruency is pursued.

4.8.2 Listen Carefully

If the situation in the cockpit allows, listening should be intensified and non-verbal signals regarded. This applies to conversations outside the cockpit, as well. It is the job of the listener and also a sign that he is "active". One should not just endure the discussion passively, in expectation that the other will be expressing himself in all clarity.

4.8.3 Specifically Question and Clarify

If issues remain that are still unclear, then these should not merely be accepted, but require clarification. Those areas for which no information is available: intentions, desires and even emotions will also be integrated in order to clarify the situation to the greatest extent possible.

4.8.4 Be Interested and Provide Confirmation

In R/T this means to "Confirm". The listener is attentive and takes an interest in the remarks of the speaker, confirming this directly and immediately as soon as a consensus exists between the two parties.

4.8.5 Repeat in Summary

The R/T readback of a clearance should serve as an example: I repeat (in my own words) what I heard and understood or what was mutually agreed upon and, in so doing, establish accord—both parties now have the same starting basis and the same mental model for the situation.

4.9 Possible Training Methods

The training content described here can be grouped into two categories: Interpersonal communication (Sects. 4.4, 4.5, 4.7 and 4.8) and technical communication (Sect. 4.6).

4.9.1 Interpersonal Communication

It has already been made clear from the content presented that this is an area you can't just "read about". Therefore, newsletters and other printed publications are not suitable for conveying knowledge in this field. They are absolutely necessary, however, as an introduction and to awaken curiosity and interest. Computer-based-training (CBT), as well, which has proven itself to be very successful for technical training applications and has met with predominantly favourable reception among the trainees, is not suited to this subject area: The knowledge of how to cope with one's own personality or how to deal with other people can not be learned with the help of a machine. In addition, it must be considered that computer programs can never adequately address or make allowance for all human behaviour patterns.

For this reason, only seminars with a limited number of participants under the supervision of trained specialists are suitable as a medium for practical training. Only personal experience can provide the insight and trigger the impulse needed to work on one's own communication proficiency. These seminars should be comprised of:

- Group exercises
- Personality assessment
- Feedback techniques
- Role playing
- Case studies
- Interpersonal experiences

The seminars, which summarize the content of several HF subject areas, should provide sufficient time for practice and for mutual criticism. Correspondingly, the use of video feedback for self-assessment will take on great significance.

A combination of seminars with subsequent opportunities for practice and repetition in a simulator-LOFT program providing video feedback should be deemed as the optimal method to this end.

CRM training is not highly regarded by all aircraft commanders (or trainers) today. Persuasive effort is needed in order to make clear the great potential for safety that is harboured in the use of these techniques. Thus, the launch of a seminar without supporting information is just not sufficient. It is absolutely essential that training in the field of Human Factors—CRM be strongly supported by the managers of the respective flight operation as well as the overall training and inspection staff. Hackman describes this process as "Altering Group Norms" (1993).

When the majority of pilots consider it as being "good airmanship" to behave according to the ideals presented in the seminars and to integrate these experiences into their daily cockpit activities, then even the sceptics will soon be won over.

> Heightened awareness will produce tangible behavioural change (Foushee 1984).

4.9.2 Technical Communication

Expert lecturers should be used in both initial as well as upgrade training. In this regard, particular reference can be made to past accidents and incidents, which may have occurred due to misunderstanding, insufficient use of standard phraseology and the careless treatment of team concepts (Crew coordination concept, Marine concept, Two-way communication, etc.). Sufficient examples can be found in the NASA "ASRS" database.

Newsletters, articles in company magazines or professional association publications can be used for a refresher between training modules.

When using LOFT's with video feedback, the effect of various communication styles and the emergence of misunderstandings can be demonstrated.

Simulator missions and check flights lend themselves nicely to the area of "radio telephony".

4.10 Error Prevention in Practice

- Speak clearly and precisely at all times (articulation and content).
- Always speak a little louder than one would normally consider appropriate.
- Make sure that you have understood correctly and have been understood correctly by your colleagues. For this reason:
  - Maintain eye contact when ever possible without distraction.
  - Listen carefully
  - Confirm messages from colleagues with a verbal or a non-verbal signal (two-way communication).
- The use of standard technical communication as prescribed in the handbook should be adhered to diligently at all times (call-outs, wording, etc.).
- If there is anything unclear or anything you don't agree with, it should be addressed immediately.
- Ambiguities must be clarified immediately and at their first indication (especially when you are still "new").
- Conflicts should be resolved as quickly and as equitably as possible. Conflicts left unresolved are a serious threat to safety.

References

EU-OPS 1.965 Subpart N (2008), Brussels

Fahlgren G (1989) Presentation about delta airlines—ALPA professional standards symposium 1, Boston

Foushee HC (1984) Dyads and triads at 35,000 feet. Am Psychol 39(8):885–893

Foushee HC, Lauber JK, Baetge MM, Acomb DB (1986) NASA technical memorandum 88322. NASA Ames Research Center, Moffett Field, CA

Hackman JR (1993) Rethinking crew resource management. In: Wiener EL, Kanki BG, Helmreich RL (eds) Cockpit resource management. Academic, New York

Hawkins F (1987) Human factors in flight. Ashgate publishing, Aldershot

Molcho S (1983) Körpersprache Mosaik, Munich

Molcho S (1988) Körpersprache als Dialog - Ganzheitliche Kommunikation in Beruf und Alltag, Mosaik, Munich

Paula M (1992) Sage, was Du meinst! mvg-Verlag, Munich

Predmore SC (1991) Microcoding of communications. In: Accident investigation: crew coordination in united 811 and united 232. NASA/University of Texas, Crew Performance Project, Austin, TX

Schulz von Thun F (1981) Miteinander reden, vol 1. Rororo, Hamburg

Schulz von Thun F (1989) Miteinander reden, vol 2. Rororo, Hamburg
Starck R (1993) VC-Human Factor-Konzept. In: VC-Info 7/93, Vereinigung Cockpit, Frankfurt
Webster's New Collegiate Dictionary (1961) Merriam webster. Springfield, MA
NASA Aviation Safety Reporting System (ASRS)

Stress

5

Hans-Joachim Ebermann and Gerhard Fahnenbruck

5.1 Introduction

It is virtually a natural part of the airline pilot's self-image to be able to adequately handle stress.

His work conditions and tasks exhibit the classic attributes of extreme stress. A high degree of responsibility, multiple burdens, time constraints, a continuously and intensively changing environment are intimately tied to this profession, just to name a few.

When stress continues over an extended period of time, it depresses performance capabilities and threatens health. According to physicians, over half of the illnesses we experience may be triggered by too much stress.

Extreme stress can lead to panic and possibly even to a loss of control over the motor functions. When individual stress limits have been exceeded, decisions can no longer be made (Koechlin 2011).

Just as with all people, pilots, too, bring stress from their private lives into the workplace, which ultimately adds to their job stress. For this reason, the following discussion is not limited to merely the occupational issues related to handling stress.

Flight-related stress plays a central role in Crew Resource Management (CRM). Among other things, decision making, information processing, human error and communication in the aircraft are directly influenced by too much, but also by too little stress.

H.-J. Ebermann (✉) · G. Fahnenbruck
Main Airport Center, Vereinigung Cockpit e. V, Unterschweinstiege 10, 60549 Frankfurt, Deutschland
e-mail: vc.ebermann@onlinehome.de

G. Fahnenbruck
e-mail: gerhard.fahnenbruck@human-factor.biz

H.-J. Ebermann and J. Scheiderer (eds.), *Human Factors on the Flight Deck*, DOI: 10.1007/978-3-642-31733-0_5,

Table 5.1 Error rates

Factor	Error rate increase
Novelty of the task	17 times
Time constraints	11 times
Too much information	6 times
Misjudgement of risk	4 times

In the course of **processing information and making decisions**, too much stress can lead to a limited capacity for receptiveness. Under some circumstances, we won't even recognize the information right in front of our eyes and have problems selecting between alternatives.

There again, decision making (or a lack thereof) is the greatest source of stress in the cockpit. The pressure associated with always needing to make adequate decisions under difficult and rarely occurring circumstances is one of the main drivers of mental stress.

The willingness to engage in risky decision making increases when time pressures become too great. When these risky decisions then lead to failures, the tendency develops to try to master the situation at any cost, even to the extent of potentially disregarding prescribed standards. The crew then very quickly finds itself entangled in the so-called **poor judgement chain**.

One indication of a crew being subjected to a high stress burden is the reduction of verbal **communication**. The breakdown of communication is one of the most commonly occurring factors in the causal chain leading to accidents.

While the frequency of **human error** *is increasing in general,* the probability of error increases significantly under stress. Williams (1988) ascertained increases in the following error rates (see Table 5.1):

Astonishingly, a systematic approach to coping with stress is usually not being taught in pilot training programs to this day. The Vereinigung Cockpit calls for the inclusion of stress training in the CRM seminars conducted during recurrent training.

The following discussion provides an introduction to the subject area of stress as well as an overview of the related content needed for airline pilot training.

5.2 What is Stress?

Stress has a cumulative effect (Jensen 1995).

Stress is the sum of all the stimuli influencing us. Eustress is positive stress, which is required for maintaining the health of the overall organism. Dystress is damaging stress, which permanently disrupts our physical and mental equilibrium and, in so doing, can damage the organism.

Stress is experienced by each person **individually**. One person may encounter eustress while the other encounters dystress in the same burdensome situation. Aerobatic flight or a CAT-III approach may be perceived as a positive challenge or as pleasure, but also as stress, depending on the individual's constitution, personality and experience. The borderline between eustress and dystress is blurred and oftentimes difficult to define.

5.2.1 Tension and Relaxation

The body is in a continuous state of alternation between activation and rest, or between tension and relaxation. The activation and tension phase is the stress phase.

When the stress phases clearly take prevalence over the relaxation phases, then physical equilibrium is no longer warranted. Dystress develops cumulatively.

Performance and rest, or tension and relaxation, are polarities, both of which belonging to a harmonious life. It is essential to establish a good balance between the two and to find "the right rhythm". If a rest period is not provided for, then tension can accumulate so extremely that it can't return back to a normal state of rest, even after work has ended. The person is then no longer able to unwind and relax. A portion of the tension is still present even at night, working to prevent him from falling asleep or sleeping soundly.

In the same manner, the balance is disturbed when the tension phases are deficient and the rest phases predominate. Here, too, negative developments can result, such as the rapid aging of older persons after completing their professional lives.

In highly automated aircraft, lapses caused by daydreaming or "spacing out" can occur during cruise flight that compromises "situational awareness" (NASA 1996).

Thus, every person goes through their own particular periods of performance-related highs and lows. This brings about very specific individual working rhythms, to the extent that external circumstances permit it. If the individual working rhythm stands in contradiction to work flow demands, then the natural process can be influenced to such an extent that signs of stress will result.

5.2.2 Stress: An Elementary Reaction

The typical stress reaction leads to changes in the body that prepare the organism for a near-term emergency reaction, also referred to as an **alarm reaction**.

In evolutionary terms, this alarm reaction serves to activate all bodily reserves for the flee or fight situations experienced by our ancestors. Even though we live much differently today, situations still arise that require our whole strength and/or attention. Imagine flying along an airway in cruise flight when an aircraft approaching from the opposite direction is recognized at the last minute as being on a collision course. A near-miss can no longer be avoided.

A typical alarm reaction would be experienced:

- All of a sudden, the person is wide awake.
- The heartbeat accelerates rapidly.
- The muscles tense up.
- The hands become moist.
- The heart feels like it's in the throat.

Unconscious changes also occur:

- The blood sugar and fatty acid levels rise.
- The blood's viscosity increases.
- The blood pressure rises.
- The metabolism changes.

All these processes are caused by the release of stress hormones.

Such a short-term stress reaction is normally completely harmless, especially when it is followed by physical activity. This is not possible, however, for a pilot sitting in the cockpit.

A bodily alarm reaction follows a stress situation. This alarm reaction activates all bodily reserves to subsequently flee or fight (with physical activity as a result). An exhaustion phase subsequent to the flight or the fight then ensues, followed by a rest period, after which the body returns back to its normal state.

This stress reaction is the organism's stereotype answer to the very diverse potential responses one can have to stress. Every stimulus to the senses, interpreted as a threat to the body or a disruption of the equilibrium, sets this in motion.

The hypothalamus, a part of the brain, triggers the release of adrenalin, noradrenalin and cortisone hormones into the bloodstream from the adrenal glands. These hormones produce the reactions just described in a fraction of a second. Any stimulus that can trigger this reaction is referred to as a stressor. Such a stressor can be acoustically stimulated, such as through an "aural warning", or even just a simple thought, such as thinking about one's own knowledge gaps just prior a check.

5.2.3 Stressors

The list of stressors is endlessly long. Just about every facet of life has triggered stress at least once in every person at some point in time.

Stressors can be subdivided into three groups:

- **Physical stressors**: noise, vibration, extreme temperatures and humidity, oxygen deficiency
- **Physiological stressors**: fatigue, poor physical fitness, illness, missed meals, low blood sugar levels
- **Psychological stressors**: mental workload, social challenges, difficult decisions, fear

Here are a few pilot-specific stressors:

Noise, low air humidity, abnormal, new airports, rushed approaches, weather, bird strikes, irregular and long working schedules, difficulties in planning free-time and social activities, as well as with maintaining friendships, responsibility for the passengers, aircraft and crew.

Stressors can also be generated through the respective airline's corporate structure.

For example, it becomes burdensome when a flight operation puts its pilots under enormous cost pressures or fails to offer stable labour conditions. At the same time, insufficient recognition and anonymity are common within organisational departments that have become too large. According to the Human Error model from Reason, these create stressors that become the *latent conditions* in later accidents.

Extraordinary personal events in life can also represent particular stressors. Pilots should carefully consider when they should return to flight duty following a divorce or the death of a close family member. The burden accompanying such thoughts lowers a person's tolerance to stress. Even simpler life situations such as home construction or financial problems can quickly impair the receptive capacity of the brain's working memory.

5.2.4 Reactions to Stress

The vulnerability to stress can be extremely diverse due to conditioning from youth (e.g. "good boy") and disposition (phlegmatic or irritable).

Anyone who was raised by his parents to be a "good boy" learned from a very early age to conform. Good performance and unobtrusive behaviour were rewarded. Poor performance and loud, aggressive behaviour were punished. The questionable result of this form of upbringing is that this child will possess an exaggerated performance consciousness even into adulthood and may have difficulty "letting go" of feelings such as anger and resentment. Such an adult can encounter problems handling stress.

In his striving for achievement, this person may be subjected to too many stressors while repressing all stress dispelling emotions.

A precise scientific analysis of this type of individual and his psychological background would go well beyond the scope of this chapter, yet, it can generally be argued that an analysis of the individual's management of stress would be necessary in order to effectively combat stress.

5.2.5 Stress and Performance

Insights into the relationship between stress and performance have been known since the beginning of the twentieth century. They are summarized in the Yerkes-Dodson Diagram.

Every task requires a specific level of stimulated arousal in order to optimally accomplish it. Too many or too few stressors degrade the ultimate work result (Beehr 1995) (see Fig. 5.1).

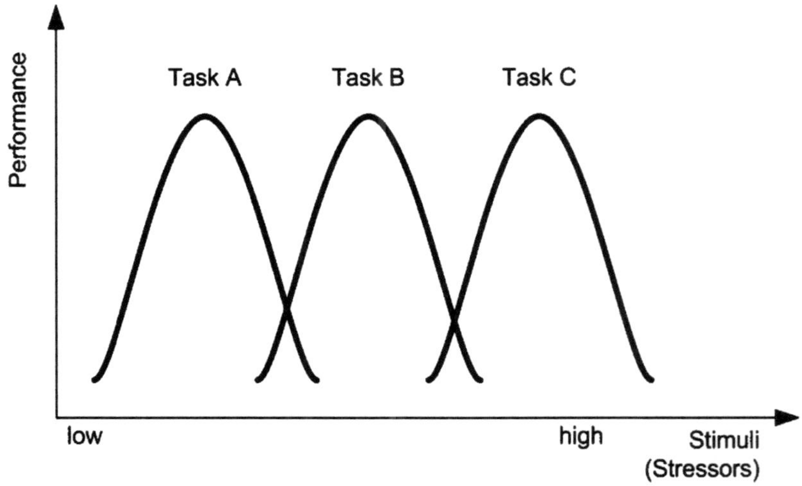

Fig. 5.1 Relationship between stimuli (or stressors) and performance

- Too few stressors:
  Boredom, fatigue, frustration and dissatisfaction
- Too many stressors:
  Inadequate problem solving, exhaustion, illness, low self-esteem
- Optimal stressors:
  Creativity, continued development, satisfaction, progress, rational problem solving

The depiction in Fig. 5.2 (Jensen 1995) has also been widely propagated:

The flight crew's resilience slowly subsides over the course of a duty shift due to fatigue. The various flights or flight phases place differing performance demands on the crew. At the same time, the crew's capacity for stress should never become exhausted.

5.2.6 Stress and Pilots

> Pilots that don't cope well with stress are more likely to be involved in aircraft accidents (Alkov et al. 1995).

Different professions, differing **stress levels**. Reference is made in this section to the English study done by Sloan and Cooper in 1986. The table below (see Table 5.2) shows the occupation-related stress ratings on a scale between zero and ten according to the estimations of stress experts (Sloan and Cooper 1986):

The airline pilot profession occupies a very high position on this list. To be a pilot means to be subjected to stress. The study also ascertained a significant

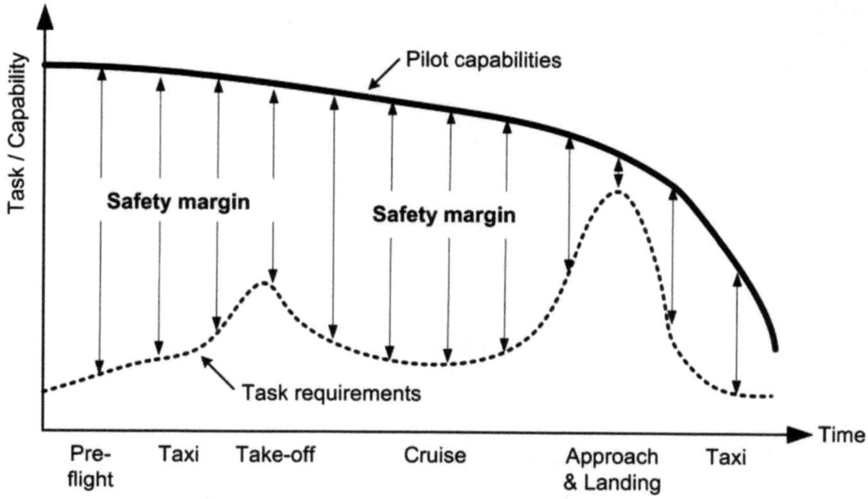

Fig. 5.2 Conceptional diagram of the margin of safety throughout a typical flight

difference between the pilot groups at 21–30 years and 41–50 years of age. Older pilots tend to be more prone to depression and psychosomatic disorders.

The pilots most susceptible to stress are those who are older, tired and weary. The same goes for pilots who do not have sufficient opportunities for rest and relaxation, and ultimately for those who cannot count on adequate support from family and friends.

Captains make up the group most particularly affected by stress. This is not due to their age, however, but because of the decisions they must make and the on board responsibilities they assume.

5.2.7 Pilot-Related Stress Model

From what has been shared to this point, a model can be built to depict the management of stress using several exemplary factors (see Fig. 5.3):

If the management of stress is inadequate, the following symptoms may develop:

- Depression
- Mental resignation
- High blood pressure
- Periods of absence
- Alcohol abuse

The symptoms arising from poor stress management function to some extent, themselves, as stressors and work to raise the already existing pressure. It's not long before one finds himself in the vicious cycle depicted below (see Fig. 5.4):

Table 5.2 Stress level related to particular professions

Profession	Stress level
Police officer	7.7
Journalist	7.5
Airline pilot	7.5
Dentist	7.3
Advertising professional	7.3
Performer	7.2
Other physicians	6.8
Med. nursing staff	6.5
Fire-fighter	6.3
Teacher	6.2
Social worker	6.0
Professional athlete	5.8
Manager	5.8
Stockbroker	5.5
Psychologist	5.2
Diplomat	4.8
Physiotherapist	3.5
Priest	3.5
Physicist	3.4
Biologist	3.0
Librarian	2.0

From what has been shared to this point, it is evident that the following points are of **considerable importance** for the cockpit working environment:

- Each crew member must actively strive to maintain a high degree of resiliency or a high threshold for dystress.
- Each crew member must be capable of recognizing the imminent approach of his dystress threshold and of responding accordingly.
- Each crew member must also be capable of recognizing the imminent approach of his own or his colleague's stress limits and be able to intervene accordingly. The working capacity of the cockpit team shall not be jeopardized.

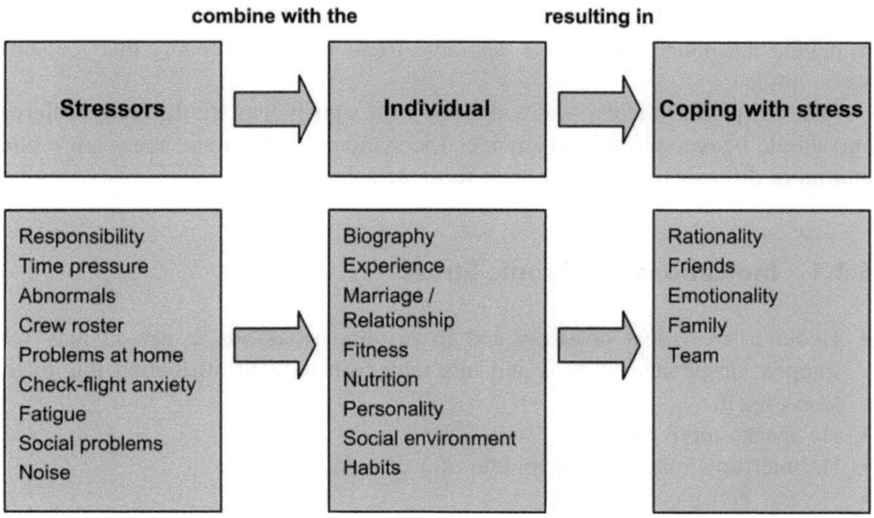

Fig. 5.3 Pilot-related stress model

Fig. 5.4 The vicious cycle of stress

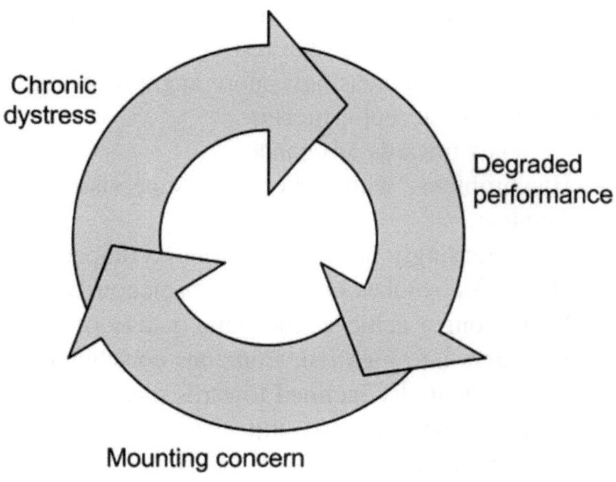

5.3 Signs of Stress

The following discussion differentiates between two types of stress: chronic stress and acute stress.

Chronic stress builds over time due to persistent stressors, oftentimes external to the profession. For the most part, acute stress results from occupationally-related stress peaks that can extend up to the limits of one's productive capabilities. Both types of stress are cumulative, adding to the total stress burden. This means that those persons already encumbered with chronic stress from home may not be able to handle that much more acute, sporadic stress before exceeding their personal stress limits.

When exposed to stress, the body sends out signals that are distinctly different and should be recognizable to anyone. The symptoms of chronic stress are subtler and more difficult to recognize than those of acute stress.

5.3.1 Indications of Chronic Stress

- Under a great deal of stress and in extreme situations, a person may feel trapped, desperate, helpless and miserable, yet without attributing this to the stress itself.
- He speaks more rapidly.
- He interrupts others in the middle of a conversation.
- He eats conspicuously fast.
- He takes on more commitments than he can successfully handle.
- He hates to squander time.
- He frequently drives too fast on the road.
- He attempts to accomplish several things simultaneously.
- He quickly becomes impatient when confronted with slower colleagues.
- He spends too little time resting or together with friends.
- He becomes increasingly more aggressive in response to criticism.
- He doesn't accept criticism.
- He tends towards arrogance.
- His problems with colleagues, supervisors or in the family become more frequent.
- He increasingly draws back from participating in social activities.
- He drinks, smokes or eats in conspicuously large amounts.
- He no longer achieves the same quality of work.
- He enters into high risk situations conspicuously often.
- He is potentially inclined towards uncontrolled fits of rage.
- He treats other persons unjustly.
- Over the long-term:
- His personality changes and he is no longer the same person he once was.
- He is increasingly inclined towards illness.
- His appearance becomes grey and pale.

5.3.2 Acute Stress

Oftentimes, several things must be taken care of simultaneously when commanding an aircraft. Demands sometimes may appear to be too great for the time available, which can lead to overstressing. A state of tension and irritation follows, along with a limited power of judgement. Headaches and digestive disorders can arise.

Time constraints and overloads manifest themselves through:

- A strained voice—rapid, high-pitched, hasty
- A tensed seating posture, oftentimes not in the centre of the seat
- Too rapid breathing rate or the holding of breath
- Perspiration
- Heavy and/or too rapid heartbeat
- Dry mouth
- Reddened skin colour
- So-called "white knuckles"
- The clenching or gritting of teeth
- Too rapid or no eye movement
- Tunnel vision
- Difficulty gathering thoughts
- Conversations will be merely technical in nature, if at all.
- No one jests or makes a joke.
- Attention is consistently drawn to incidental matters.
- Deviations from Standard Operating Procedures are undertaken.
- Decisions are no longer made.

When these symptoms appear, an increased risk for the flight, itself, exists. It is then increasingly likely that the crew will become entangled in the so-called poor judgement chain.

5.4 Poor Stress Management

> A person is responsible for himself—and only he can do something for his own inner well-being (Crisand, Lyon 1981).

Most people do not handle chronic stress very well. We have developed various and very different options for automatically dealing with stressful situations.

We can:

- Suppress the problem.
- Look to others to blame for the problem, but not to ourselves.
- Take the blame exclusively upon ourselves (Gunn and Ruthrock 1994).

Each of these options can be helpful, as long as it is not used excessively. Exaggerated usage destroys the sense of reality and therewith the objective perspective of the problem lying beneath the chronic stress. Signs of an exaggerated suppression of the problem or a too one-sided view of the problem are anxiety, irritability or depressive tendencies.

- **Anxiety:** nervousness, indisposition, sleep disorders
- **Irritability:** loss of one's own rhythm, alienation from friends and colleagues, lack of acceptance in one's surroundings

- **Depressive tendencies:** hopelessness, disappointment, inadequacy, chronic fatigue, loss of appetite, sexual disorders, loss of interest

These phenomena are oftentimes combated with alcohol. An increased rate of alcohol consumption often points to a serious crisis. Further indications that the stress management process has been disrupted can be a noticeable aggressivity, as well as any form of addictive behaviour (miscellaneous drugs, anorexia, compulsive gambling and workaholism), shopping frenzies and noticeable cynicism.

The high demands placed by pilots on their own productive capabilities and emotional stability make it particularly difficult for them to seek psychological help. For this reason, severe but treatable disorders are often needlessly protracted. Self-help groups for addiction problems have now been formed among pilots and some pilots' associations such as the German ALPA offer burnout prevention seminars. These signify the overcoming of an old fashioned pilots' self-image.

Learning to live with dystress is the worst approach to stress management. It may be unavoidable for short, manageable periods of time, but over the long-term it will create fundamental problems. An airline pilot who cannot adequately deal with stress on his own should not shy away from seeking help and advice.

5.5 Properly Coping with Chronic Stress

As a rule, the root of the chronic stresses pilots must deal with will not be found in factors directly related to flying (Crisand and Lyon 1991).

What can a person specifically do to better handle the personal stress he is subjected to?

To begin with, the management of chronic stress is an activity that must be initiated and carried out by the person, himself. In principle, there are two options to this end: The person should reduce the number and extent of his individual stressors, which is possible only to a limited degree within the scope of his profession. The other option is to change the manner in which the unavoidable stressors are dealt with.

One can obtain assistance for this by seeking the advice of a specialized psychologist, by attending a stress management seminar or by reading a book on this topic.

The methods used are approximately the same in each of these three options. A person's personal burden due to chronic stress is determined through testing. The individual's personal background and, with it, his disposition towards coping with stress are thoroughly illuminated. The main stressors for each individual are searched out and assessed. Finally, a management strategy is established together with the counsellor and progressively adapted to the current situation as necessary. In addition to these initiatives for dealing with chronic stress, there are a great many possible options for opposing the very draining professional demands through spiritual, mental and physical stability. A counterbalance to growing professional challenges, including those related to flying, can be established through leading a responsible lifestyle. The following advice can be helpful therein:

Rest and slowness are essential. One merely has to consider the natural rhythm of tension and relaxation. Those who suffer from chronic stress almost certainly enjoy periods of rest all too seldom. The novel "The Discovery of Slowness" by Sten Nadolny offers an entertaining approach to checking one's own attitude.

Friends: Especially flying personnel have frequent job-related problems cultivating and maintaining friendships. Friendships are an important counterbalance to stress.

The role of a **life partner** is particularly important: 40 % of all marriages will end in divorce if one of the partners is at the height of their professional agility but has less and less time to attend to private matters. Two thirds of all senior managers admit to having stress-related difficulties with their partners.

A pilot's life partner has an especially difficult role to fulfil. Like a single parent, they are confronted with an unshared portion of the child-rearing and household burdens due to the pilot's frequent absences. This is often a source of overburden and frustration. Moreover, pilots expect constructive support from their spouses when they return home from a flight. This same support is anticipated by the ones remaining at home, but they oftentimes don't receive it.

The partner is subjected to a further burden in that the family often finds itself in a form of isolation due to the pilot's work schedule, which doesn't always allow for regular contact with the rest of the family or friends. Persons not actively working seem to struggle with these challenges to a greater extent than their employed partners (Sloan and Cooper 1986).

What can help?

- Regular conversations with the partner are important. Problems that can't be solved alone must be openly discussed.
- Wives shouldn't altogether abandon their career plans in favour of the family. Children must be nurtured for several years; the career shapes an entire life.
- Spouses have experienced favourable results by allowing their partners to be alone for some time after they return home. The conversation can be first sought out with the question: "You look exhausted, how are you doing?" Many will grab at the chance to vent their day's affairs and talk about their experiences. Afterwards, they will be more interested in their partner's daily routine. Both will benefit once the pathway to a dialogue has been found.
- Regular opportunities should be taken to discuss those aspects of the relationship that are going well or that are not going so well.

The best and most enduring partnerships are those, in which the partners coexist as equals with one another, and not just from the other or for the other (Handelsblatt 1995).

Adequate, uninterrupted **sleep** is one of the most decisive regenerative mechanisms of the human organism. Whenever possible, sleep disorders should not be just accepted. They can be alleviated by different means. This can range from doing relaxation exercises just before going to bed to taking a short walk in the fresh air. One should avoid switching on the television and it is better to read prior to going to bed. Sporting activity shortly before going to bed is not good either, as the physical exertion temporarily acts as a stimulant (Garrett 1995).

Sleep induced by tablets or alcohol is merely a daze, but its relaxation effect is relatively low.

There has been some hope placed in melatonin, a substance produced by the brain during emerging darkness that provides the body with a "signal to fall asleep". The body is then "tricked"into believing it is nighttime. Melatonin is used to combat the negative effects of a disrupted day/night rhythm, which is typical for the airline industry. But it is not free from side effects. Among others, the feel for mental resiliency is disturbed. In addition, the body's own melatonin production rhythm can be disrupted.

Self-medication should be avoided by all means in order to prevent administering an incorrect dosage of melatonin (VC/DLR Alertness Management). Medical advice should be sought if sleeping problems remain stubborn. Perpetual sleep deficit has destabilizing effects and is a hazard to both the person's health and the conduct of a flight.

The stress burden can be reduced through **exercise**. Those who regularly take part in endurance sports are generally more emotionally balanced and physically resilient. An endurance-trained person releases fewer stress hormones when subjected to the same stress burden. The resistance to stress increases thereby. Regular endurance sporting activity is an important instrument for sensibly and successfully managing and preventing stress.

Beneficial endurance sports can be jogging, hiking, swimming, cross-country skiing, rowing and cycling.

At the same time, it is important to avoid overestimating one's own capabilities and to not carry over the performance-oriented mentality from the profession into the sport. The time invested in sporting activities should be consciously perceived as a period of relaxation following the tension built up on the job. Relaxation doesn't have anything to do with competition, confrontation or a battle against time.

Sport, as an anti-stress device, means: *long and slow*. Physicians recommend 30–60 min of training at least three times a week with a continuous pulse rate based on the empirical formula of 180 minus your age (Skolamed 1993). An exposure of 80 % of the maximum individual pulse rate would be even better.

Sensible **nutrition** is another key to both inner and outer stability. Yet, despite all the efforts to explain its significance, it is largely still disregarded. The stamina and regenerative capacity of both the body and the spirit are decisively determined by nutrition.

A balanced and regular diet is often difficult to maintain and even sometimes impossible, especially for pilots underway in an aircraft or on a layover stop. The bywords here would be a well contemplated consumption of coffee, tea and alcohol, moderation with fats, meat and sweets, foodstuffs that are as natural as possible with lots of fruits and vegetables. Aviators in particular should give conscious heed to a regular diet without hectic or distraction. A more detailed introduction to proper nutrition would go beyond the scope of this discussion, but a wealth of literature is available in any bookstore. Moreover, within the context of a

CRM seminar, an aviation physician, for instance, could be invited to lecture on the subject.

In addition, regular **relaxation exercises** are recommended to round out the program. These include:

- Progressive muscle relaxation according to Jacobsen.
- Yoga, meditation, autogenous training, as well as massages and use of the sauna.
- Long walks and other extensive outdoor activities in a natural setting are also quite simple and effective.

The success that comes with relaxation exercises will be first realized after some time. The results from autogenous training, for example, may be expected after a good six months. Therefore, some patience will be required until noticeable effects set in.

A person is very essentially the product of his own **thoughts**. In addition to physical fitness, our productive capabilities also demand mental fitness. There exists a direct correlation between thought activity in the brain and the immune system. Simply put, this means: People can either think themselves sick or think themselves healthy. Someone who is easily excitable, who can't remain at a "composed distance", who is continuously brooding and sometimes speaks to himself in negative soliloquies, undermines his long-term joy for life, productive capabilities and health. Inner stability requires more than just purpose-specific technical expertise. The mental aspects are all too often neglected by aviators. The conscious and critical search for something new when engaging art, philosophy and even religion can be helpful.

According to one study, the best way to remain healthy while overcoming burdens is to actively comprehend them as being challenges, tests and opportunities to learn something new. One should therefore confront his burdens with optimism and a sense of responsibility (Volk 1996).

Good interpersonal relationships in the everyday work routine should not be taken for granted. The thoughts invested in categories such as speed, rivalry and competition not only burden interpersonal relationships on the job, but in private life, as well. The willingness to perform and the enjoyment of work are heavily influenced by the relationship to the colleagues. It is much more pleasurable to work together with a congenial und humorous colleague. Inappropriate behaviour, on the other hand, often returns like a boomerang.

Here are a few more flyer-specific tips:

- Avoid gaps in the theoretical knowledge needed for flying. Gaps cause insecurity.
- Consciously plan the use of time. It is better to arrive ten minutes earlier at the dispatch office during poor weather in order to avoid time pressures from the outset due to possible alternate planning difficulties.
- Make deliberate use of every training opportunity, including simulator events and line checks. Checks serve as a further training opportunity just as do performance-level tests. Good training creates self-confidence.
- Avoid the regular consumption of alcohol, such as might be common upon completion of a fight. Alcohol is of no benefit whatsoever when it comes to

combating stress over the long-term. The threshold to addiction is closer then most people think.

- It may be better not to report for flight duty when under a high degree of mental strain, perhaps due to a severe personal experience.
- Cockpit colleagues should be openly and earnestly informed about burdensome factors such as a sleep deficit, household problems, illness, etc. The colleagues' reservations about addressing unclear work processes will be reduced thereby.
- A colleague under stress will commit more work-related errors than normal. The error chain can be broken at the first sign of error through explicit and immediate enquiry on the part of the colleagues in the cockpit (Lufthansa 1994).

5.6 Properly Coping with Acute Stress

Never be in a hurry. If you are in a hurry, you are in danger (Aero Safe 1991).

Acute stress is the sum of all sporadically acting stimuli.

Examples might be an approach that is progressing too high and too fast, a hasty prepared flight due to a delay, marginal weather conditions, an abnormal procedure or a base check.

The probability of error increases significantly when time pressures exist. An analysis done in America regarding incidents where time pressures were indicated as being the triggering factor led to some interesting conclusions (McElhatton and Drew 1994):

- In 65 % of all incidents, the causal errors were committed by crew members who had a mental or emotional disposition towards being in a rush.
- Most time constraint-related errors were made during the flight preparation phase, followed by the taxi-out and take-off phases.
- The study offers the following advice for avoiding error when time constraints exist:
- Beware of a **rushed and hectic pace**, especially during flight preparation and taxi-out. Pilots must work to counter any external distractions (ramp agent, station, passengers) or pressures for punctuality when they arise during these phases.
- Be sensitive to the pressures as they start to build up. More time should be explicitly and purposely taken at these times in order to work through the pending tasks according to the urgency assigned them.
- Any non-essential tasks should be put off until a later point in time.
- Checklists should be read through completely and correctly at all times.
- When a procedure or a checklist is interrupted, it should be read through again from the beginning.

Here are a few more tips:

Naturally, acute stress can also be associated with all other work phases. In stress situations, one should be fully cognizant of his physical rhythm: demand, performance, fatigue and subsequent rest. He should take advantage of short periods of rest as soon as possible in order to be prepared for the next demand as it arises. Beware of the poor judgement chain. When subjected to stress, the probability of error increases, so that when an error is made, the level of stress rises even further and the next error will occur that much sooner.

Accept one's own flaws. Respond to every error, even those you may have just made, yourself.

When subjected to stress, one tends to deviate from the SOPs. It is precisely in such situations that they should be complied with as faithfully as possible with no allowance be made for one's own, or a colleague's deviations. Stress situations should be deliberately and actively rectified whenever possible. This could mean initiating a go-around, entering a holding pattern, rejecting a "line-up and wait" authorisation or simply waiting a few moments.

Rectification should also be sought when it is noticed that one of the crew members is "not in the loop". Flying represents teamwork in practice and is not a stage for a "one-man show".

When a burdensome situation cannot be avoided despite these options, at least they can be better controlled through the appropriate and timely application of the following "emergency measures":

Relaxation Through Tension

The body's entire muscular system should be tensed and then subsequently relaxed. In so doing, a pleasant heaviness in the muscles will be noticeable. The relaxation of these muscles results in an inner calmness.

Conscious Breathing

The breath becomes quicker and flatter when subjected to acute stress. As a countermeasure, one should breathe in deeply and slowly through the nose while counting to five. The breath should then be held for three to five seconds and subsequently let out slowly while counting to five. The tension will noticeably dissipate from the body when this is repeated several times (Skolamed 1993).

Good experience has been made with autogenous feedback training. A group of pilots undergoing this training subsequently exhibited better results in the simulator (Kellar et al. 1993).

Endurance sports immediately following the flight will be beneficial for acute stress in the same manner as with chronic stress. As already mentioned when discussing the "elementary reaction", all bodily reserves should be mobilized to flee or to fight. The physical activity that follows then helps return the body to a normal hormonal balance. Endurance sports following a flight thus become our contemporary replacement for appropriately dealing with stressors.

5.7 Stress Following Particularly Burdensome Events

Events can occur while flying that go far beyond the extent of what a person in the given situation can cope with. In this case, it is not so much the event itself that matters, but the perception of the person affected. A good example would be the crew member who, upon assessing the inability to retract the landing gear due to a technical defect as a well-manageable task, is ultimately capable of landing the aircraft safely. Another crew member, in contrast, may potentially recall scenes of collapsed landing gears and fear for his life right up to the landing.

Although seemingly astonishing at first glance, it makes absolutely no difference whether the person is male or female, has a lot of experience or just a little, or sits on the right side or the left side of the cockpit. Just how the event is ultimately handled is clearly not dependent upon these factors.

It almost seems as if it is simply a matter of who gets caught when, and by which situation. Literally everyone can be affected.

Nevertheless, if it is assumed that the event, itself, is the trigger and not how it is dealt with, then the experiences gained through the event become critical, especially if the life or soundness of one's self or a person in the immediate vicinity was threatened or appeared to be so.

Classic examples are:

- The death or near death of a passenger on board
- Even more critical is the death of a colleague on board or during a layover
- The threat to one's own life and limb, such as might be caused by:
- system failure (in particular an engine or a landing gear)
- events based on external conditions, such as wind shear, turbulence or thunderstorm

The event, in this context, is not the problem, but rather the threat to life and limb that is perceived as arising from it. The risk of inappropriately coping with the situation is particularly great when the event occurs suddenly and unexpectedly.

Even though the resistance to stress and its management are enhanced through the techniques already mentioned, such as endurance sports, autogenous training, yoga, etc., everyone can still get caught off guard just the same.

Of course, not every event will leave a long-term impact on those involved. On the contrary, critical events in approximately 80 % of all cases will be handled by the effected persons alone and without any difficulties. 20 % of those affected will react more intensely. These reactions can be mental, physical or emotional in nature and be expressed though conspicuous behaviour. As a general rule, these are normal symptoms in healthy persons subjected to abnormal situations. The symptoms may be unpleasant for the effected persons at first, but they will prove to be of no real significance or not even bad. Only in extremely rare cases or when they persist longer than four weeks will they need to be treated. If crews do not receive support following these so-called "critical incidents", however, approximately 4 % will begin to suffer from depression, substance abuse or posttraumatic stress disorder (PTSD).

A list of the normal reactions and options for coping with them can be found at the homepage of "Stiftung Mayday"(www.Stiftung-Mayday.de) under Downloads, "CISM-Informationsbroschüre" and "Belastungshinweise" (the "CISM-Critical incident stress management information brochure" and the "Instructions regarding burdensome events" are both available in the German language only at the present time).

Stiftung Mayday has made it its goal to support license holders and their dependants in the German-speaking regions during and following critical situations. This support has resulted in a reduction in the number of long-term crew member illnesses from 4 % down to 0.8 %. At the same time, the progression of the illnesses that still arose proved to be much milder in scope and shorter in duration because of the support. The number of near-term illnesses has thus been reduced by 80 %. The work carried out by Stiftung Mayday is absolutely confidential and is not conducted inside the airlines intentionally, even though an airline may support the initiative both organisationally and financially. Crews should be provided an opportunity to talk about their situations within a framework that has nothing to do with the disciplinary structure. In the first instance, this is good for the effected crew members, but it is also of benefit to the employers because the rate of illness is lowered and the duration of those illnesses is shortened.

An "operational debriefing" should be carried out following any event that a crew member deems to have been a critical incident. The compilation of all views will contribute to a much more coherent overall picture as well as to considerable relief.

Operational Debriefing

When an abnormal incident leads to differing perceptions within the effected crew, the commander should then conduct an "operational debriefing" with all crew members.

Such an abnormal incident is defined according to three criteria:

- It takes place outside the scope of the daily routine.
- It leads to a deviation from the normal work routine.
- Differences are likely in how the incident may have been perceived and assessed.

The operational debriefing is confined to the pure facts. It should be carried out as soon as possible following the incident. Non-effected persons are not permitted to participate or to listen in. Examples could be: a go-around, a diversion to another airport, etc.

The following questions should be addressed and clarified during the operational debriefing:

- What are the facts and what course of action was followed throughout the incident?
- Are there differing perceptions among the crew members?
- Is there a need for continuing CISM measures?
- How should we proceed from here?

CISM measures are necessary when an operational debriefing lasts longer than 15 min and strong physical or emotional reactions are recognizable (see Fig. 5.5).

When an operational debriefing is not sufficient and the crew members still require further support or simply have open questions, they can contact the Stiftung Mayday hotline at any time: +49 (700) 7700 7703. Initial contact at this telephone number will be with a call centre employee who will record the call back number and a short description of the incident. Call centre employees are trained merely to gather information about and to record an incident. They are trained to

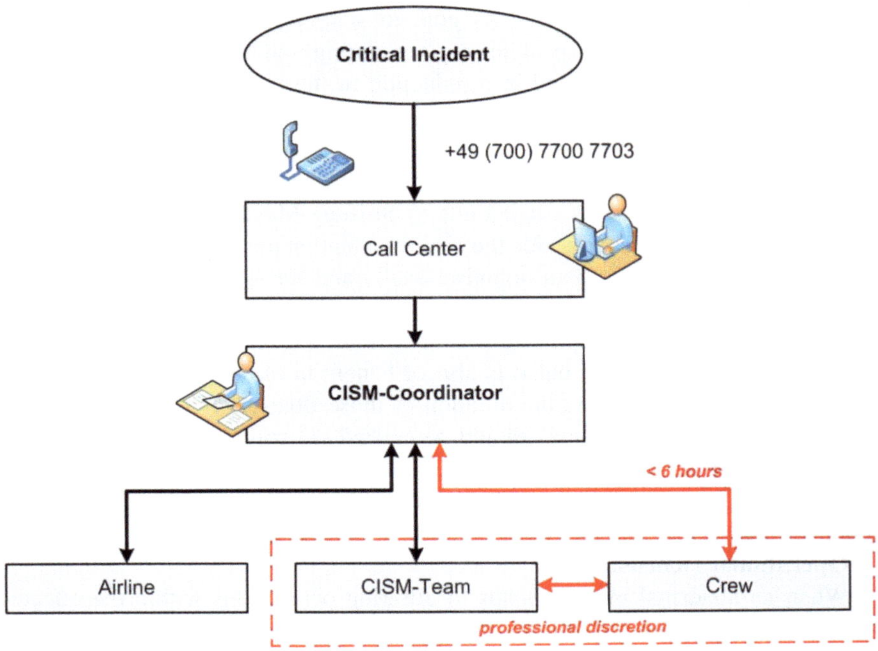

Fig. 5.5 Intervention following a critical incident (according to Stiftung Mayday)

provide a straightforward telephone support service, then to forward the telephone number with a description of the incident to a coordinator who will return the call to the initiator within 6 h. The caller can be an effected crew member, himself, or another person altogether. This service is also provided in English.

Coordinators are key figures in the concept of crew member support following critical incidents. They are very well trained colleagues out of the flying profession (cockpit and cabin) and have a great deal of experience. They will discuss the event and initiate measures as necessary—always in agreement with the caller. Such measures can be the conversation, itself, a conversation with a specialist, a pickup at the aircraft once it arrives at its base airport, a briefing with the crew immediately following its return or a few days afterwards. In some cases, information is initially sent via e-mail or fax in order to arrange a further conversation. No matter how the ultimate measure is carried out, the conversation with the coordinator as well as any subsequent conversations with other trained colleagues from the CISM team are bound by a commitment to professional confidentiality through Stiftung Mayday.

Stiftung Mayday is able to arrange for further measures at the request of the individual, but only if this process is not sufficiently helpful and he takes ill following the incident. The goal in every case is to ensure the full recovery of the effected colleague as quickly as possible.

5.8 Trainability

Stress is a popular subject; one that has encouraged a wealth of general literature. The ability to successfully cope with stress is obviously of significant importance for airline pilots.

Yet, one would think just the opposite is the case, as there is surprisingly little scientific material available addressing the effectivity of stress management training (Beehr 1995). The effects of the individual measures, such as relaxation training and biofeedback among others, are minimal or simply not verifiable by themselves. Furthermore, stress training sessions are aimed for the most part at the individual, while studies related to anti-dystress measures in teams or at the company level are virtually non-existent.

Thus, the knowledge available demands a holistic approach to addressing the stress potential of the individual. Only through comprehensive knowledge about stress and the recognition of how it manifests itself, along with the acquisition and the application of good stress management techniques, can its negative consequences be prevented.

With this in mind, we consider the following training content/learning objectives to be essential for attainment of the airline transport pilot license:

- Definition of the terms Stress and Stressors
- An understanding of the individual reactions to stress
- An understanding of the correlation between stress and performance
- An understanding of pilot-specific issues related to the subject of stress
- Recognition of chronic and acute stress
- An understanding of unreflective and poor stress management
- An understanding of, and training about how to properly cope with chronic and acute stress, with a focus on clear communication regarding stress limiting factors relevant to the cockpit crew, with a deliberate and concerted effort to avoid haste and hectic in the preparation and conduct of a flight. Moreover, the targeted intervention into the development of error chains and the active rectification of stress with a sensibility for crew workload can all be trained.

The knowledge referenced above with a focus on the practical aspects are also essential components of CRM seminars that should be repeated on a regular basis.

These practical training points should be continuously taught and refreshed throughout the professional career as part of the practical training that accompanies it in the simulator and aircraft.

We consider the conception of a pilot-specific seminar that comprehensively deals with chronic and acute stress to be highly desirable.

All newly initiated instructional and training measures should be accompanied scientifically and be examined for their effectivity on all accounts.

5.9 Relaxation Techniques and Addresses

Respiratory Therapy

Respiratory therapy with simple exercises leads to a resumption of the natural breathing process and to the dissipation of tension. Ilse Middendorf provides instruction in her book "The Perceptible Breath: A Breathing Science", Junfermann Publishing House, with further addresses and links at the German language website for the "Arbeits—und Forschungsgemeinschaft für Atempflege e. V."(Research foundation for respiratory care).

Autogenous Training

This method of relaxation instils a feeling of heaviness and warmth through concentration focused on the body. Health insurance companies should be able to provide information about competent trainers.

Meditation

Meditation focuses on a form of mental release or "switching off" rather than on physical relaxation. It functions through concentration on the respiratory rhythm, on music or through complete silence. An introduction into Zen is offered by the Shido Centre in Worpswede, Germany.

Yoga

Yoga functions through a tensioning and relaxing of the limbs in alternating body positions.

References

Aero Safe (1991) VC-Info 7/91, Vereinigung Cockpit, Frankfurt

Alkov et al (1995) In: Matthew: Aviation, space and environmental medicine (01/1995)

Beehr (1995) Stress in the workplace. Routledge, London

Crisand E, Lyon U (1981) Anti-Stress Training, Arbeitshefte Führungspsychologie, Sauer

Crisand, Lyon (1991) Anti-Stress Training, Arbeitshefte Führungspsychologie (management psychology workbook), Sauer

Garrett (1995) MedAire Inc., USA

Gunn, Ruthrock (1994) in Air Line Pilot (06/1994), U.S. ALPA

Handelsblatt (1995) from 4 Dec

Jensen (1995) Pilot judgment, Avebury, UK/USA

Kellar et al (1993) In: Flight safety digest 7/93, flight safety foundation, Washington

Koechlin (2011) Lecture at the DGAC Congress, November 2011, Paris

Lufthansa (1994) in part from CF-Info 2/94, Dept. FRA CF, Frankfurt a.M

McElhatton, Drew (1994) Air Line Pilot 8/94, U.S. ALPA

NASA/Langley Research Center (1996), In aviation week and space technology 8 April 1996

Skolamed (1993) Fit zum Führen, Holzmann

Sloan SJ, Cooper CL (1986) Pilots under Stress. Routledge, London

Stiftung Mayday (2012) www.stiftung-mayday.de

Volk (1996) In: Handelsblatt from 1 Mar 1996

Williams J (1988) A data-based method for assessing and reducing human error to improve operational performance. In: Hagen W (ed) IEEE fourth conference on human factors and power plants. Institute of Electrical and Electronic Engineers, New York

Decision Making

6

Johannes Bühler, Hans-Joachim Ebermann, Florian Hamm and Dagmar Reuter-Leahr

6.1 Description of the Problem

Virtually 50 % of all accidents are related to "poor airmanship" or can be traced back to faulty decisions and even a lack of decision making by the respective crews (Lufthansa 1993).

In a study made of turbojet aircraft accidents in the USA compiled by the National Transportation Safety Board (NTSB), the following facts were determined (NTSB 1994):

- "Faulty or improper decision making" was the primary cause in 47 % of the total losses examined.
- Faulty tactical decisions were made in 67 % of all accidents. This could mean, for example, the failure to make a decision despite the presence of clear action signals, as well as the non-observance of warning or alert signals.
- In 40 % of all accidents, the faulty decisions made by the captain were not challenged by the FO. This is often the case with a decision to forgo an otherwise mandated go-around procedure.

J. Bühler (✉) · H.-J. Ebermann · F. Hamm · D. Reuter-Leahr
Main Airport Center, Vereinigung Cockpit e. V,
Unterschweinstiege 10, 60549 Frankfurt, Germany
e-mail: johannesbuehler@t-online.de

H.-J. Ebermann
e-mail: vc.ebermann@onlinehome.de

F. Hamm
e-mail: florian.hamm@lft.dlh.de

D. Reuter-Leahr
e-mail: reuter@emotions2lead.com

H.-J. Ebermann and J. Scheiderer (eds.), *Human Factors on the Flight Deck*,
DOI: 10.1007/978-3-642-31733-0_6, © Springer-Verlag Berlin Heidelberg 2013

With respect to the last point, Boeing determined that 14 % of all accidents to-date would not have occurred if a timely decision to go-around had been made (Boeing Commercial Airplane Group 1993).

Consequently, poor decision making, or the failure thereof, plays a substantial role in the cause of accidents. Why is this?

One possible reason could certainly be the role-change that pilots have undergone up to this point in time. In the early days of commercial aviation, pilots were seen primarily as "craftsmen" but, with the introduction of semi-automatic flight control, they have now become "system managers". In today's environment of sophisticated automation, highly complex systems and artificial intelligence, the pilot is increasingly cast as the "strategist" and "decision maker" at the centre of the flight process. Correspondingly, his capacity for decision making is of increasing importance.

Furthermore, those areas of pilot responsibility have been and will continue to be expanded due to the elimination of cockpit and ground personnel (radio operators, navigators, flight engineers and on-board mechanics, as well as station personnel who handle catering, boarding irregularities, etc.). The pilot must now take on an increasing number of tasks that he would have previously delegated and assigned the responsibility to somebody else. This complicates his task and increases his workload, meaning that timely and well-founded decisions are more difficult to make.

Today, the pilot must coordinate all operational procedures in an extremely limited period of time (e.g. during transit times or on short-haul flights). The routines developed for this are dominated by automatisms. These run almost reflex-like and relieve the pilot of the continuous concern over every minute detail—he does not need to establish new plans (= make deliberate or elaborate decisions) because he has already made them at some previous point in time. He has developed very specific methods from experience and is of the opinion that, using these, he can handle just about any problem that may arise. Particularly when these methods have proven themselves in practice over time, an overestimation of their effectiveness can set in (Dörner 1996). The routine application of action patterns will usually lead to burden relief, but this approach proves unsuccessful when complex decisions must be made or unfamiliar correlations must be dealt with.

The following sections are meant to illustrate these decision routines, as well as to portray the detailed, deliberate decision making process. What should a pilot know about them? What mistakes can be made in the process? How can he defend himself against making them?

6.2 Definitions

If the term "decision" is to be defined, it must first be distinguished from the terms "decision situation" and "decision making". A decision situation is a prerequisite for a decision. Decision making is the method by which the decision is reached.

Decision Situation

A decision situation is present when multiple action alternatives exist.

Decision Making

Decision making is the mental process of collecting and analysing all available information in a specific situation, as well as the deliberate weighing of action alternatives leading to a timely decision as to which course of action to take (FAA 1991).

Decision

A decision is the choice of one of the possible action alternatives.

Uncertainty

Decisions are made under conditions of certainty or uncertainty (Hanf 1986). As a rule, when certainty exists with regard to the outcome of two action alternatives, the decision is made easy because the better alternative is oftentimes obvious. Decisions made under uncertainty are difficult. In the process, it is absolutely essential to understand how one can incorporate uncertainty.

The outcome of the action alternatives must be forecasted and can never be clear-cut. In principle, predictions are uncertain and *subjective* because they are influenced by individual experiences and emotions.

Uncertainty will always result because it cannot be precisely known just how much value the data underlying the decision will have. The acquisition of essential related knowledge takes time.

Uncertainty causes stress: the need to make a decision is a significant stressor in itself. High levels of stress, in turn, negatively impact the quality of the decisions made.

In summary, a decision made in a complex and time-critical environment, as is commonly the case in aviation, can be defined as follows:

- A person (group of persons) chooses only one course of action from at least two action alternatives.
- The choice of a course of action takes place on the basis of the expected results from the various action alternatives.
- The choice is made on the precept that the action alternative selected is subjectively perceived as promising the greatest advantage. This definition assumes that the decision maker has a precise preference with regard to the result.

6.3 Characteristics of the Decision Making Process

6.3.1 Time Available for a Decision

The decisions made during normal flight operations take place along the line in the graph below (see Fig. 6.1):

If only minimal time is available, decisions will be made rapidly and intuitively. The decision maker uses a spontaneous reflex, automatism or a so-called heurism.

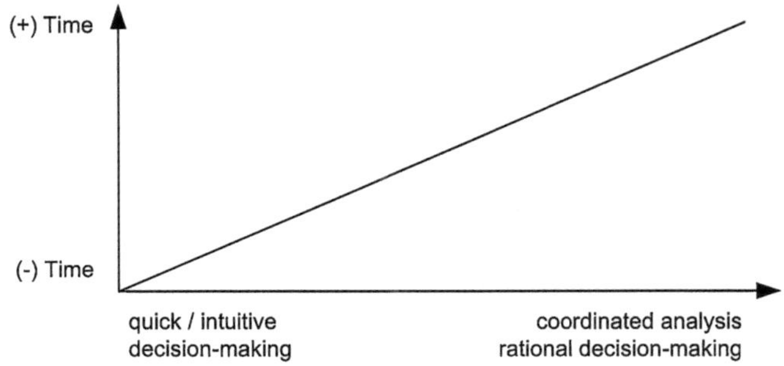

Fig. 6.1 Correlation between time available and the type of decision making

A heurism is understood to be a situation that has already been played through or is familiar, in which one mostly reacts and decides automatically. This helps to minimize the stress associated with time constraints. It is also referred to as Naturalistic Decision Making (NDM). Having already played through, or being familiar with a situation means that the decision maker is either well prepared, well trained or possesses a great deal of experience. With growing experience, the decision maker will be able to apply NDM more often, thus avoiding the more complex decision making processes that will be described later in the text.

The decision maker must also consider his colleagues throughout the process in order to avoid making decisions on his own. These colleagues must then be decidedly incorporated, step by step, if they are not able to use NDM on their own in the respective situation due to lack of experience. This may take time, but it improves the accuracy of the decision.

6.3.2 The Complexity of a Problem

The simpler a decision is, the more the decision maker will tend to rely on impulsive reflexes or on heuristic NDM. When they become more complex, however, he should pursue a coordinated or an analytical decision (see Fig. 6.2). At the same time, "simple" and "complex" are relative: The experience and the knowledge of the decision maker are decisive. An experienced check captain can draw from a considerably broader band of situations and knowledge than a newly trained entrant into the profession, for whom a comparatively simple problem from the check captain's perspective may prove to be a highly complex one.

Accident research reveals that there have been problems with both the "simple" as well as with the "complex" decisions.

It is important that flight crews be given the tools necessary to define a coordinated,

- analytical decision making process and to systematically identify
- the processes that—particularly when under stress—impede optimal decision making.

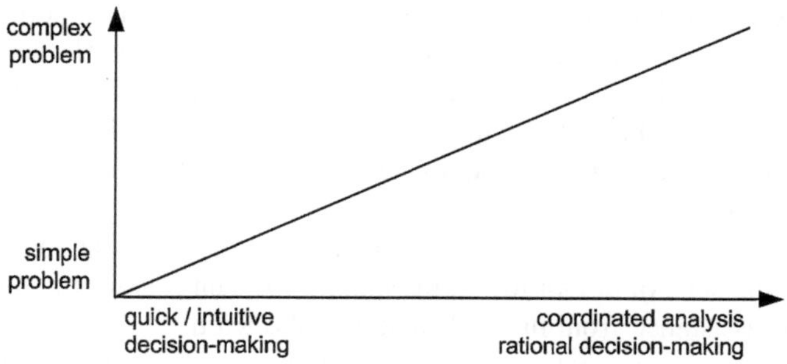

Fig. 6.2 Correlation between complexity and the type of decision making

- Furthermore, they must know what influencing factors they are subjected to, particularly with respect to making quick decisions.

The initial discussion that follows addresses the "simple" decisions. Thereafter, the focus will be on the more infrequent "complex" decisions; those that must be consciously thought through and coordinated, and for which adequate time is available.

The decision making process will be difficult to impossible if the flight crew, and above all the captain, is improperly positioned with respect to the axes on this diagram, meaning that too much or too little time is taken for a problem, or that the analytical decision is dealt with in too shallow or in too detailed a manner.

6.4 Quick, Intuitive Decisions

6.4.1 Preliminary Note

Whether or not a decision is made according to rational considerations is dependent on the perception of the person affected. He is more or less capable of drawing on his relevant experiences from memory, depending on what pictures related to the current problem form in his mind and which past experiences, training sessions and feelings are coupled with them.

The better trained and/or the more experienced a pilot is, the quicker he will be able to make a decision. In so doing, he can call up conscious knowledge that can be retrieved within seconds because of routine or practice.

Studies show that unconscious knowledge can also be retrieved. New experiences make it possible for it to adapt to the changes in demand. Unconscious knowledge is called into action when the problem doesn't correspond to previous experience, when the risk exceeds all prior conceptions or when time pressures don't permit any further deliberation. As opposed to a state of not knowing, unconscious knowledge is stored in the brain because the affected person may have once heard of or witnessed the experiences made by others.

The state of not knowing, on the other hand, is greater than both the unconscious and conscious knowledge combined. This usually reveals itself after the fact, when one realizes with surprise that he merely supposed himself to be well versed in a matter. The state of not knowing is substantiated only through very limited potential foresight, so it requires a high degree of imagination and a good sense of intuition.

The more complex things get, the greater the state of not knowing becomes. The greater the state of not knowing, the more important it is for the pilot to admit his need for assistance from his colleague and the team.

Access to one's own intuition and the routine in dealing with it are indispensable, especially in moments of crisis, in order to determine in the shortest possible time just what is and what is not known.

For this reason, intuition, or unconscious intelligence as scientists have designated it, offers a beneficial complement to the decision making process, thereby enhancing the rational thought process.

Our decisions are not rational because, with the exception of standards (e.g. SOPs), experience, emotions and rewards all play a role in the process, as well (Elger and Schwarz 2009).

6.4.2 Scientific Background

Intuitive capacity, which had been shunned for years, was ascribed new significance in the 1980s. The scientific disciplines of psychology, medicine, mathematics and philosophy followed up on the insights gathered over the course of history and began attempting to combine the available models into an interdisciplinary concept. The following insights are presently being compiled in research literature.

Unconscious perception and information processing

As early as 1910, Poincaré was referring to the "incubation", the submerging of the mind into the unconscious that, after some activity, would lead to an intuitive solution to a problem (Landua 2009).

The German psychologist, Gerd Gigerenzer, director of the Max Planck Institute for Human Development in Berlin, presently stands out as a successful researcher in the field of unconscious intelligence. He traces perceived knowledge back to the human system of perception (Gigenrenzer 2007). Our eye is not sufficiently capable of recognizing reality without any shadow of a doubt. Researchers from the CERN Centre even claim that the eye can only perceive 5 % of the total spectrum surrounding the person. Thus, 95 % of reality would not be ascertainable for the observer. Our brain is thus challenged to fill in the related "gaps" that emerge with plausible correlations, thereby coming to unconscious perceptive conclusions while calling upon past experiences. In so doing, "gut feelings" quickly emerge in the consciousness. We don't understood completely why we have these, but we are prepared to act upon them. These feelings usually arise unconsciously. They can be heightened at the conscious level, however, through active training. We can perceive what the eye wants to see and what the feelings sense.

Intuition through practical knowledge

Practical knowledge is an accumulation of information units, consciously and unconsciously gathered over many years, which the decision maker can resort to very quickly. Albeit, these do not refer to specific, concrete facts stored in memory, meaning they do not deal with the recollection of a specific earlier experience. The experienced person draws rather from a pool of numerous similar situations, which have combined together over time into a vast wealth of experiences. This practical knowledge exercises considerable influence over one's view of a situation, which is generally beneficial. Notwithstanding, it can turn negative if a person has been doing something wrong for years, yet still calls it experience. An example of this might be the long-lasting disregard for an SOP, which misleads the individual to view this SOP as not being safety-relevant. The fact that the SOP may have evolved from just one accident means that his "practical knowledge" is erroneous and misguided.

The more information available in the pool of experience, the more quickly and accurately the decision maker will be able to respond. Nevertheless, it is incumbent upon his perceptive abilities to always weigh the value of this experience anew. Example: The experienced pilot above who disregarded the SOP for years should be able to perceive his mistake and, there upon, commit himself to comply with it in the future.

Moreover, practical knowledge is subjected to continuous stimuli from outside sources. According to expert estimates, we take in almost 220,000 times more information units unconsciously than we do consciously. At first, the brain is not capable of distinguishing whether the information from the outside world is beneficial or negative for the decision maker. In addition to the common sensory stimuli such as noise, rapid movement, light, etc., this has to do with the moods of other people in the surrounding environment (see mirror neurons) that inadvertently effect one's own thoughts and, in so doing, one's unconscious actions.

He who expands his perception through training is capable of clearly comprehending this correlation. Instantaneous actions can be better controlled and a conscious selection of the available information can be made.

Somatic markers

Rational decision making completely excludes feelings. It takes time, especially with unknown situations, because they demand a comparison of pros and cons, and result in the decision maker butting up against the limits of his attention and memory capacity in a jungle of mental branching.

If the decision maker intends to incorporate his feelings into the thought process, he creates another opportunity for limiting the decision's manifold possibilities.

Antonio Damasio uses the term "somatic markers" to point out how the sensations of the body (*soma* means body in Greek) mark certain mental images (Damasio 1994).

These body "markers" ostensibly represent an automatic warning signal that says "caution, there is danger ahead if you choose this option that can lead to this result".

With respect to the decision making process, somatic markers are important because they can draw attention to past events that are not related to the current response options. If the actor is conscious of his behaviour, however, then he can undertake a goal oriented selection. It is incumbent upon him to distinguish whether these bodily signals will have a beneficial effect or a hindering effect at the moment the decision is made.

Consider the example of a person who felt at school somewhat uneasy when giving a presentation or reciting a poem and this feeling was reinforced by the remarks of his classmates. Similar feelings may still exist up to the present day as an adult and impair one's self-confident demeanour and fluent speech, such as during a cabin briefing.

This instrument should be considered merely as one element of the intuitive capabilities and can not suffice for the normal decision making process. It is rather an additional aid in the decision making process that the decision maker can reference for rechecking whether his feelings have anything to do with the immediate situation.

The fractal affect logic

Similar to Damasio's thesis, the research findings from Ciompi are based upon the assumption that thoughts of any kind are closely tied to behavioural programs and feelings (Ciompi 2002). This system continues over an entire lifetime, so that the branching out of our system of association becomes continuously more complex.

External stimuli trigger affects in the organism. According to Ciompi, these will always arouse feelings, emotions or moods that are physically perceptible. At the same time, he even refers to the body as one's own organ of feelings. These feelings exhibit the brain's electrical manifestations and, for this reason, are graphically ascertainable by means of EEG equipment. Using this method of measurement, it has been possible to prove that human being is *permanently* subjected to the influence of these affects so that he is constantly emotionally attuned.

Conclusion: without emotions there is no thinking. They are responsible for our focus of attention, link the thought content from memory storage and reduce the complexity of our perceptive content. Because of this, colours, for instance, will be perceived as being gloomier and thoughts will become more destructive than usual when feelings of sorrow prevail.

In the sense of self-similarity (fractality), Ciompi refers to the impact emotions have on the bodily processes and on the behaviour patterns of the person concerned at the same time. This explains why fear, for example, can manifest itself in a cramped physical posture and timid behaviour.

In the interest of cockpit safety, therefore, it is essential for the pilot to learn how to monitor his own thoughts. He should take a critical, in-depth look at his own behaviour patterns and consistently realign his perceptions to his body. If he can recognize his typical behaviour patterns and is capable of viewing his feelings and his behaviour from different perspectives, then it will be possible for him to transform a portion of his unconscious responses in daily life into conscious action.

Example: A person senses a sudden occurrence of heat or begins to perspire (body signals). This is possibly an indication of stress. The affected person recognizes this warning signal and is aware that he is momentarily at risk of making more mistakes. Or: a colleague acts or communicates in such a manner that anger or even rage is triggered in others (unkempt appearance, poor discipline, etc.). When one senses these emotions, he first has an opportunity to dispel them within his own self (perhaps the colleague is not aware of the effect of his behaviour) in order to be subsequently able to address the critical behaviour in a more congenial manner.

If the pilot is aware of the effect of affect logic and is capable of communicating his observations of his team member's behaviour patterns, then the areas of responsibility related to a colleague's suddenly emerging, obstructive emotions can be promptly shifted.

Example: The FO recognizes that the captain is taking too much time to solve a technical problem at the parking position and offers to perform an outside check.

The abdominal brain

The abdominal brain, also referred to as the Enteric Nervous System (ENS), is a mesh of nerve cells encompassing the entire intestinal tract. In humans, it possesses four to five times more neurons than the spinal cord. The neurobiologist who discovered it, Michael Gershon, views this "second brain" as being a copy of the cerebral brain. Cell types, active substances and receptors are exactly identical. It functions independently of the Central Nervous System (CNS) and can sense independently (Gershon 2001).

Hence, decisions based on "gut instinct" are made not due to clouded reason, but emanate from a perception through the intestinal tract that takes place considerably more rapidly than one through the brain. This is based on the realization that, as with a dedicated line, the abdominal brain continuously sends the greater portion of its perceptions to the cerebral brain where the information units are immediately assessed by the limbic system, which is the part of the brain responsible for processing emotions.

The superiority of the abdominal brain, resulting from the speed with which it can transmit information, leads to the conclusion that its messages will always be sent prior to the rational explanations. In so doing, it provides an elegant reference in preparation of that, which the cerebral brain will shortly be transferring and, as in all other scientific explanations for intuition, represents an ideal supplement to direct rational knowledge.

Giving heed to the signals is of great importance. This can lead to an enhancement of assertiveness, for instance, if it is done consciously and taken seriously. Uneasy feelings should be addressed because, as a rule, they are justified. Early processing can enhance safety and reduce subsequent stress.

We sometimes refer to this as an aviator's seventh sense. An example could be an uneasy gut feeling on the part of the Pilot Monitoring when the Pilot Flying is trying to "push through" an unstabilized approach. Addressing the situation at an early stage with strong assertiveness, which can lead to a go-around, increases safety and reduces stress.

Mirror neurons

Mirror neurons are nerve cells that trigger the same potential in the brain when a process is being viewed as would be generated if the person were carrying out the process on his own. They become active when an activity is performed on one's own, but also when observing a person performing the activity or even when merely imagining the activity.

These automatically functioning, mirroring and imitation responses serve to support the development of a small child. With the aid of mirroring phenomena, a child copies every movement his mother makes and imitates it in order to acquire a new skill. An example of this would be the mother opening her mouth when feeding a small child so that the child will move his mouth to take in food.

The system of mirror neurons is also present in adults. It enables us to experience and empathize with the feelings of others. With its help, one can simulate actions by way of these feelings and better understand the intent of outside actions so that a possible outcome can be anticipated.

The mirroring capacity, or resonant capacity, is operative independent of a person's age. The more contact with feelings and communication one experiences in his childhood, the better prepared he will be as an adult to put himself in the situation and the emotional world of others, as well as to reflect on their perspectives.

Joachim Bauer, a German doctor for internal medicine, psychiatry and psychotherapy, studied the biological changes triggered through the mirroring of feelings and through these also affect the brain. If certain signals are regularly produced by our surrounding environment, then not only will behaviour conform, but so will the neurobiological events in the human brain, as well. This means that the human brain is influenced, among other things, by speech and it also explains why emotional and response behaviour is trainable through conditioning of the proper speech.

The pilot can take advantage of the effect of mirror neurons if he trains his intuitive perceptive abilities. If he has learned to consciously steer his own behaviour and to specifically target the language he uses, then he can make a significant contribution to the cockpit environment and, with this, to the overall safety. Especially pilots with little practical knowledge have found this form of intuition to be helpful in support of rationality.

In this manner, it may be easier to recognize whether a colleague is in good shape or poor shape, for example, or whether he is actually certain of his decision at a critical moment. A person can learn to understand how to follow his own perceptions and how to use the moment of inner uncertainty in order to reach a point of joint, resource- and team-oriented action.

6.4.3 Limitations of Conscious Information Processing

Even though engineering technology strives to simplify the workflow in the cockpit, the number of problems to be solved grows progressively with the increasing scope of pilot responsibilities. The cockpit team is confronted with an ever expanding flood of information. And it is safe to assume that the complexity of the demands has not yet reached its limits, especially when considering the example of the steadily increasing air traffic density.

The conscious mind is able to handle only a small amount of information, achieving around seven inputs simultaneously. It becomes overburdened and breaks down as soon as this is exceeded. It is at this point that the usual form of analysis within the decision process begins to cloud our judgement. The variety and complexity of the information creates a state of not knowing in the pilot. He is confronted with what is for him an unfamiliar situation; one that he has neither passively nor actively seen; neither heard about nor experienced. His learned workflow process routines reach their limits, yet he must confront these suddenly occurring situations without any prior knowledge. At that moment, his only option is to fill the gap left by his state of not knowing with intuition.

To be sure, the pilot's brain is capable of partially suppressing the flood of information and, through a division of labour in the cockpit, of limiting his concentration to only a few inputs, insofar as both the complexity of the problem and time allow. But if the stress becomes increasingly greater, however, the emotional arousal within the problem solver will increase and his decision will either be made unconsciously or not at all.

Intuition is seldom logical because a person's emotional world is not designed to travel down a structured, linear pathway. The absence of logic and the fact that it is still a relatively young science are among the reasons why intuition, while recognized in private life as a relevant medium, still raises questions in the fields of science and technology. Yet, the knowledge gained to date regarding the emotional world of the human being gained to date is reason enough to establish a place for intuition in the basic and advanced training of pilots. Emotion and reason are not adversaries. Much more, they work hand in hand and can be complementary in more than just critical situations.

Example: Empirically, the failure to make a go-around decision is one of the most frequent decision-related errors. Cockpit crew members can reach their cognitive limits if they find themselves in a high and fast approach, no matter what the reason (air traffic control, weather, etc.). The pace of aircraft guidance suddenly becomes very rapid and complex. The adherence to descent rate and speed limits, the compliance with obstacle clearance heights, the extension of flaps and landing gear and the communication with the air traffic controllers can all overburden the brain's capacity for assimilating information. The pilot may perceive bodily signals in such a situation, such as a temperature rise in the head or body, moist hands or perhaps even speech problems. When these bodily signals are perceived and the pilot is trained to break off the approach and start it anew in response to the signals, then he has taken a major step towards safety.

6.4.4 Limitations and Dangers of Intuition

It is the fundamental nature of the brain to view its own concept of reality as always being correct, irrespective of whether it is successful and useful or not. The more faith a person has developed in the success achieved through his own intuition, the more certain he will be in dealing with the sensations and reality confirmed through them. At the same time, it is precisely therein where the danger lies.

Even though a problem may already show clear indications of dissimilarities in the general framework conditions, it is likely that particularly experienced decision makers will suppress some details and tend to rely on their previous successes. Moreover, if the situation is very complex while rational understanding has reached its limits, trivialization can emerge and the decision will be made without any further networking to other sources.

Intuition through practical knowledge alone is not sufficient at this moment because, even if it was valid yesterday, it may not fit into the general framework conditions under the circumstances today.

If stressors impede the decision maker from gathering all the necessary information, interfere with his capacity for clarifying things or completely distract his attention away from the current task, then rationality is close to retreating into the background (tunnel vision). It is advisable at such times to discuss the matter of one's own intuition together with the colleague. It is very risky to give heed to the intuition of only one individual, as it is not known from which experience base or from what emotions it may have emerged.

But even when intuition emerges collectively, it has yet to be proven just how close it approximates real systems. For this reason, it tentatively certain that intuition serves exclusively as a *supplemental* assessment, while success through intuitive action without rationality should for the time being still be considered a miracle.

6.4.5 The Personality of the Individual

The emotions

"80 % of all decisions in critical moments are made within one minute" (Klein 2003).

Moreover, if the decision comprises a high degree of risk from the perspective of the decision maker, then emotions and particularly fears are aroused that make it difficult to control the mind. Emotions change the mental activity as well as the physiological state (e.g. heart throbbing, blushing) of the person. This is because there are many more nerve connections directing information from the centre of emotion to the centre of understanding than the other way around. For this reason, feelings exercise considerably greater influence than reason.

Thus, when decision moments arise suddenly, emotions can induce an uncontrolled interpretation of the situation in the affected person. They influence his behaviour and, in so doing, his environment.

What's more, pilots unconsciously share their emotions among themselves without recognizing that their actions are reciprocally influenced by their own fears. Even if the behaviour seems at first to be as normal as always, the initial thoughts arising will be co-determined by each individual's physical arousal within seconds while a "dialogue of inner voices" begins.

At this moment, it is important to be aware of a methodology, through which the process that has been working unconsciously up to this point can be comprehended. If the feelings are elevated into the consciousness instead of being suppressed, this works already to alleviated their dominant effects and their consequences can be better managed. Just the perception of the situation, itself, can be enough to set in motion a directional change in the automatic behaviour pattern. The better a pilot has learned to understand his own fears and frustrations, and the better he is presently able to accept the behaviour of his colleagues without subjective assessment, the more rapidly he will be able to deal with abruptly ensuing emotions in a solution-oriented manner and, in so doing, limit the time required for a conscious acceptance of the emotion.

The narrower the window of perception is, the greater the probability that behaviour will spin out of control due to emotions such as fear or anger (see Fig. 6.3).

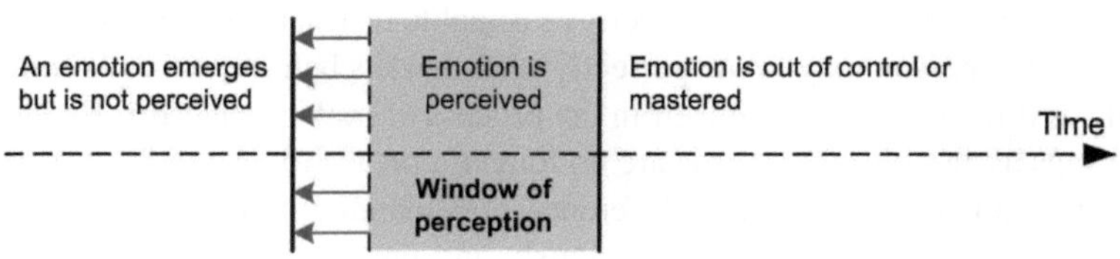

Fig. 6.3 Window of perception

A pilot's goal must be to broaden his window of perception to the left, in line with the depiction above. In practice, this means that he must be aware of his behaviour patterns and his fundamental fears. Only then can they be recognized at the moment they appear and be relativised as a consequence. The pilot is then in a position to respond rapidly and more appropriately with greater composure through this form of conscious perception, as opposed to responding reactively and, thus, unconsciously following his habitual mechanisms. Experiments have shown that the emotional reactions will weaken when they are put into words (Elger and Schwarz 2009).

The period of time between event and perception will be shortened as a result of awareness training, meaning that conscious action can be employed earlier on.

What influences will an emotion be subject to?

A pilot's emotions have a significant influence on his actions. These emotions will be determined by the intensity of the external stimuli and his personal experiences, as well as by his relationship to his surroundings and the interacting emotions they contribute.

Hence, there are many components that need to be trained up front in order to enable the pilot to control his behaviour, even at the moment of the emotion.

- He should know which intense emotional experiences from his past influence his present actions.
- He should understand what is required to create an optimal working environment from the beginning of the flight on.
- He should be capable of recognizing his own emotions and of bringing these under control.
- He should be able to recognize emotions arising in his colleague early on and should be aware of what he can do to contribute to the colleague's objective ability to actively solve the problem.

Personal behaviour patterns

Every adult person is more or less familiar with those personal traps he steps into time and again, especially as a result of emotional arousal. It happens quite often, for example, that one will reject the expectations of others with increasing emotional intensity the more these are understood as being demands, irrespective of whether they are seen as being good or bad.

This pattern of behaviour could be accounted for in that the person may have developed a mechanism as a child that permitted cooperation only in conjunction with recognition. If this same person as an adult is simply issued a demand, yet totally without an element of recognition, it is likely that he will automatically and, thus, unconsciously reject even perfectly reasonable instructions from his colleague.

Although a person may know perfectly well that his behaviour is not conducive, he feels as if he is watching himself in the process of making an error. Even though his feelings assure him that words are no longer a means to an end, he nevertheless seems incapable of behaving any differently and abandoning his uncalled for pride.

Behaviour patterns that were consciously and unconsciously acquired over the years, and which were once used to protect the "inner" child, such as those that developed in the childhood home, can prove to be devastating today, especially in relationships with others. The disregard for a colleague in the cockpit can have serious consequences because this will impair the capacity of both human resources to act.

The active confrontation with one's own thoughts and actions that draw attention to the "typical" behaviour in a moment of crisis can thus be beneficial.

If, for example, problems were not openly discussed in the childhood home, the child will act similarly at first later on in life. If the person is aware of this or is perhaps made aware of it by others during a training session, however, then he can discard this particular behaviour pattern through appropriate training and begin to open up to problems and conflicts.

Then there's the "good boy effect": Was the person raised in an authoritarian household where his youthful resistance to his parent's wishes was frowned upon and often punished? If he is not capable of quickly and clearly challenging the captain as an FO, then an accident scenario can rapidly develop. An awareness of this personal behaviour pattern can greatly facilitate the necessary critical faculties and, thereupon, the conscious cultivation of them.

Perception

The more a pilot knows about his "typical" behaviour patterns, the sooner he will be in a position to monitor himself and to correct his actions accordingly. Every behaviour pattern is accompanied by more or less discernible emotions.

Road traffic provides a good example: Someone is tailgating a driver who falls into the behaviour pattern of trying to defend his position in the "pecking order" on the highway. "He'll just have to wait; I'll drive the way I want to drive". A slight tap on the brakes will teach him to keep his distance.

If this person doesn't recognize the connection between his behaviour and the emotion, he'll commonly try to avert this emotion and seek justification in the actions of the other: "He shouldn't be tailgating me".

Perception, on the other hand, means that he is cognizant of his emotions and recognizes them as his own. A pilot trained to discern emotions knows how to separate them from abruptly occurring stimuli within seconds and to concentrate on the objective issues. He knows how these stimuli typically effect unconscious changes inside him, but knows too how to prevent this from happening.

Therefore: "All right!! So he's tailgating me. That's his problem. I can only drive as traffic conditions allow. I'll let him pass the next chance I get".

6.4.6 Training of Personality and Applied Intuition

The human is essentially a social being. His mind is deeply rooted in the social life, which is why it is so difficult for him to draw back into his personal inner life. If he wants to elevate his own degree of influence on the events around him, however, then he must be consciously aware of his own personality. Only after he has learned to understand himself and is in a position to continuously reflect on his own behaviour will he then be in a position to activate and apply his intuition; a seldom used resource.

The improvement of flight safety through a person's behaviour while flying assumes the following:

- The pilot is interested in his own potential course of action.
- He takes part in a training initiative aimed at improving his perception of himself and his surroundings. This will facilitate his ability to put these theoretical approaches into practice.

Training for the enhancement of Naturalistic Decision Making

- Acquire knowledge about the typical, often habitually recurring personal thought and behaviour patterns.
- Recognize and steer one's own emotions.
- Develop a consciousness for the continuous influence one's own behaviour patterns have on others.
- Focus perception on the body and its feelings.
- Learn to deal with the images of past stress moments.
- Intuitively determine the moods of those in the immediate surroundings.
- Know the effects and applications of "correct speech" as a positive influencing factor for an optimal cockpit environment.

At the present time, this material is being imparted solely in the context of the "VC seminar for leadership and teamwork".

Day-to-day training in the daily routine

Regular reflection on this training material is part of a pilot's continuous professional development, even in everyday life (family, relationships, road traffic).

Sustainability cannot be achieved through training alone. It is rather the application of simple tools aligned to one's daily private and professional life that makes preventive efforts that much more effective and successful. These include:

- The search for images and metaphors that sharpen both internal and external perception.
- Rituals that facilitate the training of one's own daily thoughts and actions.
- Simple speech exercises that alter the thought process.
- Practicing that relaxation that directs attention towards the bodily feelings.

6.4.7 Rewards

In addition to experience and emotions, rewards also play an important role in the intuitive decision making process.

This can be the reason why unstabilized approaches and long landings are not aborted or why diversions made necessary due to miserable weather are not initiated.

These pose little problem in the simulator, but the situation is totally different in practice. Because, if a flight is continued despite such limits being exceeded, the reward will be a timely completion of that flight. Difficult piloting manoeuvres associated with the go-around will be avoided along with the loss of face in front of the colleagues and the passenger announcement.

If the diversion can be avoided, the reward will come through the completion of flight duty. This is especially true of long flight duty times, where an extension of flight time, heightened fatigue, difficult operating conditions at the alternate airport, a commander's decision to prolong the maximum duty time, etc. can all be avoided.

The rewards for violating aeronautical limits can be tempting but, in many cases, they have marked the path to the accident. Pilots should be aware of this fact, accepting and using the process as a plumb line to separate safe actions from unsafe actions.

6.5 Decision Making Process

Having described quick, intuitive decision making to this point, the following discussion addresses its other extreme: the structured decision making process; how decisions should be made in complex situations where sufficient time is available for a decision.

This process requires a simple and structured model with which even complex decisions with a high degree of difficulty can be dealt with (see Fig. 6.4).

A multitude of decision models exist, of which the most important ones are introduced in the annex.

Within the German aviation industry, the FORDEC Model has gained prominence since the 1990s (Eißfeldt et al. 1994).

Pilots are sometimes of the opinion that they don't need a decision model at all. Kathleen Mosier determined in a study that many experienced pilots don't apply decision models because they use "non-analytical, yet very efficient detection techniques and aligned selection procedures" (heuristics) (Mosier 1991). As described above, however, this is optimal only for decisions that are either relatively simple or time-critical.

6.5.1 Situation Analysis, Fact Finding

Before the actual decision making process can begin, the situation must first be analysed for the following factors:

- Is there a need for action?
- How big of a threat is the problem?
- How complex is the problem?

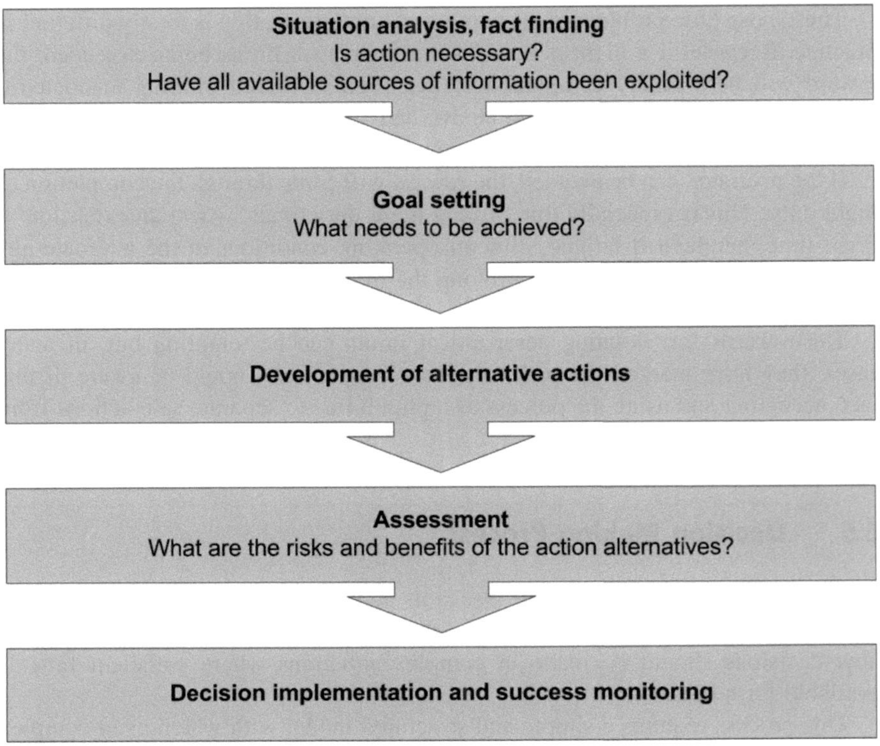

Fig. 6.4 Flow model for complex decisions

- How much time is available?
- Have all sensible sources of information been used?
- What do we know?

A need for action is given when the achievement of an existing goal is in jeopardy or an existing goal needs to be replaced by a new goal.

Example: The pilot arrives at his destination airport after a long-haul flight with a minimum of extra fuel. The destination airport is experiencing bad weather to the point that a successful landing is not necessarily guaranteed and some approach delays, in particular, should be expected. A "commitment to stay" stands in opposition to a diverted landing at an alternate airport, with imminent duty time violations if the flight is resumed from the alternate back to the destination airport.

Or: The aircraft has a hydraulic problem. Landing flaps and the braking system are available only to a limited extent. There are two possible airports to choose from: One has relatively good weather yet the runway is short, while the other has a very long runway available but the weather is poor.

It is quite obvious that a need for action is given in both examples portrayed. In practice, however, there will be a whole series of situations that aren't so obvious, such as during ground transit times where the cockpit crew takes an active part in the events. Although a catering reorder may serve the satisfaction of the customer, yet the resulting delay will achieve just the opposite. It is fairly clear that criteria other than the recognized need for action must be considered.

One criterion is how threatening a situation actually is. The threat related to a problem will determine the priorities ultimately established. Every pilot is familiar with the maxim that has accompanied him throughout his career: "Fly the aircraft first!" The fact that this principle is so self-evident makes it that much more incomprehensible when it is breached time and again. If several problems arise simultaneously, then even a good solution to a subordinate problem will not defuse the overall situation if a more important or more urgent problem is neglected. Going back to the example described above with the action situation occurring during a transit time: A potential departure delay would probably be acceptable depending on the degree of motivation and workload of the cockpit crew, but not if this would jeopardize the overall departure due to the loss of a slot time, however.

A priority is easily established in the example above. The situation is clear and unambiguous.

The situation would be different, however, if the problem or the overall situation were to become more complex. What would happen if the attributes of an action situation were not directly accessible to, or could not be determined at all by the decision maker (non-transparent)?

Example: A technical problem is encountered while taxiing to a runway at a foreign airport. A technical point of contact is not available, the MEL (Minimum Equipment List) can be ignored following "Off blocks" according to the definition of the flight operation and the technical board documentation is kept very brief. Thus, a decision must be made without any background information.

A further trait of a complex action situation is that they are usually comprised of many variables that are linked together and more or less influenced by one another. This means that the solution to one problem may be bring with it an entire series of resultant problems.

Example: Following the long-haul flight mentioned earlier with the weather and fuel problems at the destination airport, the crew decides to initiate a diversion. This may solve the original problem, but they are now confronted with a wide range of new problems: flight duty time violations, handling at the alternate airport, connecting flights for the passengers, possible cancellation of the following mission, etc.

The situation gets even more drastic when it gains momentum and takes on a life of its own. This necessarily leads to the question: How much time is available? Time pressures can arise both due to the problem itself, as well as due to the fact that the situation is dynamic in itself. In some cases, it may be necessary to refrain from collecting all the potential information available because the process of gathering it will collide with the action demanded under the time constraints existing.

Telfer uses the term "Judgement" to summarize the key aspects of the situation analysis described here:

> Judgement is an intellectual process, through which pilots identify, analyse and assess information about themselves, their aircraft and the operating environment in such a manner that a timely decision can be made that contributes to flight safety (Telfer 1989).

Finally, the question should always be posed as to whether all reasonable sources of information have been tapped. Have all the handbooks been referenced within the timeframe available? Could air traffic control or the crews from other aircraft have some helpful information? Is advice available through the various departments within one's own company? How about the cabin crew or the passengers? Is there a competent expert on board, perhaps among the passengers?

6.5.2 Goal Setting: What Needs to be Achieved?

Before starting to develop action alternatives, it should first be clear, just what is it that is actually to be achieved. Certainly, there are distinct situations where the goal is so clear that this question doesn't need to be raised at all, such as with an engine fire. Yet, in everyday practice, crews will more frequently be confronted with situations that aren't so clear and where a specific goal must first be established. Moreover, it would hardly be possible to develop a precise strategy for action by merely establishing a global objective. An example of a global objective could be "everything needs to improve" or "I want to be on time".

It should also be noted that several goals can be pursued at the same time. These can be mutually exclusive, which would mean they would be solved only through compromise. An example of this would be an attempt to pursue "the best possible quality at the lowest possible cost".

Aside from the goals already mentioned, implicit goals must also be considered. These are goals that someone may not even know he has. They oftentimes play an unconscious and emotional role in every decision, such as experiences, emotional needs and moods.

It is also important that a person doesn't attempt to rectify a state of deficiency (negative goals) through avoidance goals, but Pursues the desired state through positive goals.

Example: It is awkward to undertake not flying in a northerly direction due to the mountains. It is rather better to say: "I'll fly in a southerly direction where it is flat".

At first glance, one might say this is splitting hairs because the end result will be the same; namely that a southerly direction will ultimately be flown. Unfortunately, this is not the case. The reason lies in the nature of the human thought process.

The human imagination does not know negation. This means there is a danger that, under stress, the "don't fly the northerly route" will become "fly the northerly route".

Negative goals should thus be avoided and positive goals should be established. In precisely the same manner, the setting of global objectives should be avoided and specific goals should be targeted. Only a tangible goal will result in a tangible result. Orienting one's self merely on the blanket requirements of the moment will lead to a lack of concept, the possibility of solving the wrong problem and to uncertainty (FAA 1991).

6.5.3 Development of Action Alternatives

Every goal-oriented decision making process is based on the structure of defined and clearly distinguished action alternatives. Only in this manner is it possible to weigh the opportunities against the risks during the later assessment of action alternatives. The action alternatives should be consciously and rationally developed. A possible danger exists in the development of both too few as well as too many action alternatives depending on how much time is available, the urgency of the situation and the complexity of the problem. Too few action alternatives can be avoided by initially listing even the absurd options, just as is done when brainstorming. Too many action alternatives in time-critical situations (e.g. smoke, fire) can be avoided by limiting consideration to what are clearly the most important goals.

6.5.4 Assessment

Following the analysis of the situation, establishment of the goal and development of possible action alternatives, a forecast and assessment of the eligible action options is undertaken.

The assessment of action alternatives can first begin when it is certain that all the information required for a selection has been obtained. Too little information can lead to errors in the decision making process, while too much information can lead to possible uncertainty. The opportunities and risks associated with the possible action alternatives must now be weighed. Priority must be placed on the effective elimination of safety risks, only after which should the economical and operational issues be considered.

6.5.5 Decision Implementation and Success Monitoring

Action follows the decision: The action alternative selected is implemented and it must be accompanied by continual success monitoring.

- Are the requisites for the action still valid?
- Have the expected results actually been realized?

With regard to success monitoring, it is important to know that people give up ground only reluctantly once they have taken up a position. Consistency is seen as a virtue, meaning that many people would be prepared to maintain the status quo, even if it means accepting a higher degree of risk than with another possible option. As a rule, once someone has made a decision, they will stand by it.

Success monitoring is a reliable option for optimizing actions as they are taking place. It is a fact that the best crews made their decisions before they were in possession of the latest available information. Nevertheless, they used the subsequent information upon receipt to check their decisions and to change them as needed (Klein 2003).

Every decision, no matter how well planned and implemented, conceals a certain degree of residual risk, totally independent of the fact that each decision making process is subject to disruption in itself:

> Because of the randomness of nature, decisions made optimally will sometimes lead to undesirable outcomes (Bohlman 1979).

6.5.6 FORDEC Model

This elaborate description of the complex decision making process has been transcribed into a well-established acronym known as the FORDEC Model (see Fig. 6.5):

6.6 Disruptive Factors and Behaviour

Active factors in the decision making process can be visualized with the help of the following diagram (see Fig. 6.6):

6.6.1 Environment

It is characteristic of aviation-related decisions that they are embedded within a periphery of factors that are continuously and dynamically changing in themselves, as well as in their interaction among themselves. As such, the crew has a problem in that they have little or no influence on the starting variables. The situation changes continuously and, under some circumstances, a decision already made may require revision even while it is being implemented. This generates time pressures and contributes to the complexity of the decision situation.

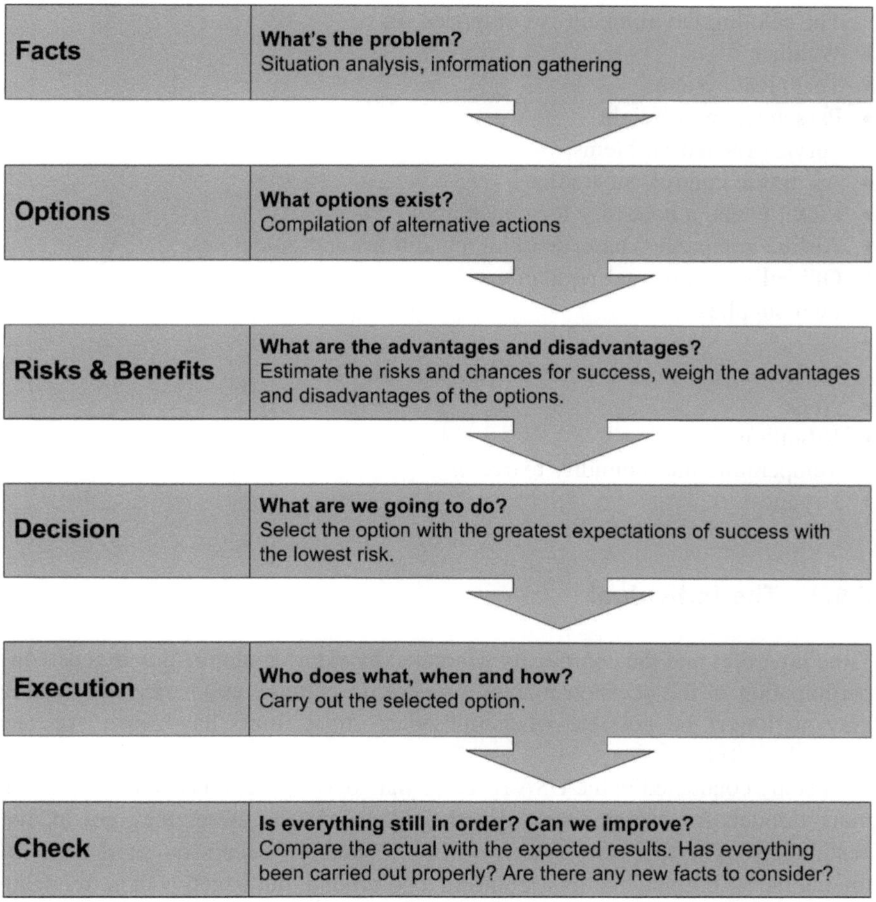

Fig. 6.5 FORDEC Model

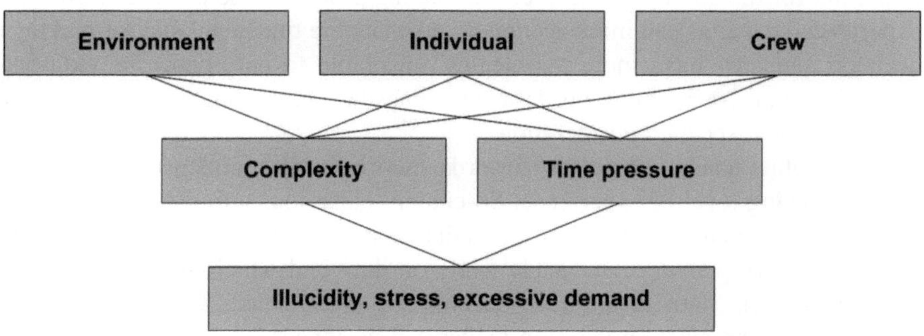

Fig. 6.6 Active structure in the decision making process

The piloting environment is comprised of:

- Weather
- Technical systems
- Passengers and freight
- Service-related problems
- Air traffic control, air traffic
- Flight plans, a necessity for punctuality
- Airline companies, basic working conditions and guidelines
- Official governmental regulations
- Fuel supplies
- …

Moreover, physical and mental stressors play a role:

- Noise
- Vibration
- Temperature and humidity extremes
- Fatigue

6.6.2 The Individual

Time pressures and the complexity generated by external factors impair a person's participation in the decision making process. In extreme cases, decision making may no longer be possible when individual stress limits have been exceeded (Koechlin 2011).

A study conducted in the USA revealed that, once made, it became increasingly more difficult for captains to revise their decisions the closer they got to their destinations. This becomes extreme when a situation occurs on short final following the "Continue" callout requiring a go-around, but which is then frequently ignored (TSB 2007). Numerous accidents originated precisely during this situation.

The experience of a responsible decision maker, meaning the captain, is important for the quality of the decision making process. Studies have shown that experienced captains can more accurately estimate the time available for making a decision and they have more "strategies" available to help generate additional time (Flin et al. 2008). Young or less experienced captains should be aware of this fact and give it special consideration.

On the other hand, a propensity towards uncritical self-confidence can also arise with a growing level of experience. So-called "hazardous attitudes" (FAA 2008) can develop that have a direct impact on the decisions made. These are personality traits that every person bears inside while on duty and which are pronounced to varying degrees. If these attitudes become too pronounced, then the individual pilot's duty as an aviation decision maker can no longer be tolerated.

Anti-authoritarian: The disregard for and deviations from Standard Operating Procedures ("SOPs are for the others, but not for me!") and/or regulations.

Impulsivity: Premature action without prior situation analysis, the lack of consideration for alternatives and/or uncalled for, instinctive action.

Faith in one's own **invulnerability**: The feeling of absolute certainty, sovereignty and freedom from anxiety. This is a phenomenon predominantly found among very young pilots with just a few years of professional experience, as well as in older, very experienced pilots.

Self-overestimation: Self-overestimation and arrogance—the compulsion to demonstrate just how good one is. The entertaining of unnecessary risk.

Resignation: Thoughtless subordination and the compulsion "to not attract any attention". Too great a desire to seek compromises due to a lack of self confidence. Can be an indication of "mental resignation".

Complacency: Overemphasis on the routine. Decreasing readiness to stay in practice. Superficial work habits, deficient advanced planning, inconsequence.

Exaggerated consideration: Age differences, authority, rhetoric, group pressure and customer demands all exercise a great deal of influence on pilots.

Psychological aptitude tests generally serve to identify persons with a pronounced susceptibility toward hazardous attitudes so that they will not be admitted into pilot training. Not withstanding, hazardous attitudes can also appear from time to time in experienced pilots, usually having been evoked by external influences and experiences the person would have recently been exposed to. It is expected of the pilot that he will learn to recognize these attitudes and, when they do emerge, to minimize them accordingly in order to keep their detrimental effects as slight as possible.

6.6.3 The Crew

In addition to the individual disruptive factors mentioned above, other factors can arise that influence the crew as a team. The most significant detrimental effects related to decision making lie in the hierarchical gradient and in communication. Both can disrupt the transmission and use of important information (Klein 2003). These are:

Centralization of authority

A definite rank order or hierarchy is formed when authority is centralized. The overall competence, including decision making powers, is transferred to the group superior. The superior gives only little consideration to the advice of his colleagues; the colleagues accept this and, as a result, offer less assistance and even hold back information (Diskell and Sallas 1991).

Weak leadership on the part of the captain

Weak leadership prevents the development of critical loyalty within the crew. Too great a degree of harmony impedes the mutual exchange of professional criticism. A trend develops towards committee decision making.

Pilot flying

A study compiled by Orasanu reveals that, during a complex decision making process, it is better for the responsible aircraft commander not to act as Pilot flying so that all his available capacity can be directed toward that process. Therefore, it is expedient for him to transfer control of the aircraft to the FO during the decision making process and then to reassume it later, as necessary (TSB 2007).

Hierarchy within the group

Differences related to rank have a greater influence on the decision making structure than the influences related to stress (Diskell and Sallas 1991???). The higher ranking person places greater emphasis on the advice of a lower ranking person during tense situations than during normal situations. With respect to their views when going through the FORDEC model, subordinates, when under stress, will tend to more readily accept the decisions made by their supervisors without criticism. For this reason, the upper ranking person should initially allow the lower ranking person to share his ideas at every step of the FORDEC model.

Too steep a hierarchical gradient can result from a large difference in age or experience, the position within the respective company (check pilot, executive) or through one-sidedly superior rhetorical abilities. A captain must be able to refrain from exercising his dominance if the performance of his crew will be impaired by it. An important prerequisite for this is the appropriate feedback from the crew.

Situational awareness of the group

Group situational awareness is not the sum of the situational awareness of the individual crew members. Research findings reveal that the "situational awareness of the captain" is the all-determining factor (Schwartz 1989).

***Tendency towards* "double thinking"**

If someone does something, the other person will do something similar at the same time. This behaviour can be very often observed in the cockpit: If one pilot replaces his approach charts, the other pilot will begin replacing his charts as well.

Fatigue

A study conducted in Australia determined that fatigued crews made decisions more slowly and were more prepared to take risks than rested crews (Petrilli et al. 2006).

Inadequate communication

The required communication within the group is heavily curtailed when conflicts remain unresolved or the level of distress becomes too great. More detailed information can be found in the respective chapters.

6.6.4 Complexity, Stress and Excessive Demand

Decision making processes are often endangered. The more unclear and complex a situation is, the more important it is for the crew to adhere to a decision model. An endangered decision making process can be restarted at any time. At first glance, this statement may seem to collide with the fact that, on the one hand, the decision must now be made in an even shorter period of time, yet, on the other hand, it is essential to avoid the risk of another rushed (wrong) decision so the crew does not find itself in the "poor judgement chain" described in the chapter on "Human error".

All disruptive factors provoked by the surrounding environment, the individual or the crew lead to an increased level of complexity, which can further lead to more or less rapidly ensuing stress, overburden and fear of failure. The point at which this occurs varies from pilot to pilot and from situation to situation. It is also not necessary to attempt to define a generally applicable entry threshold. It is much more important that each individual learns to recognize the symptoms that indicate an onset of stress for him personally.

6.7 Summary

- As a rule, decisions are made in uncertainty.
- Uncertainty leads to stress.
- Decisions are made between two poles: quick/intuitive and coordinated/rational. The more one decision model is applied, the more time will tend to be required.
- The more complex a situation appears to be, the more essential it is to use a decision model.
- As a rule, previous experience is very beneficial. Nevertheless, it should be willingly subjected to review for currency and its influence corrected as necessary.
- Armchair flying, mental training and attentive simulator training all support the formation of heuristics.
- When subjected to time constraints or in situations without practical knowledge, "gut feelings" should be listened to. If someone has an uneasy feeling about a decision that he can't rationally explain, it should be listened to and another option pursued.
- One's own feelings should be monitored and their influence used when making decisions. The "window of perception" should be consciously broadened.

- Negative personal behaviour patterns should be known and a conscious effort made to oppose them.
- Short-term rewards can be dangerous if they mislead a person into violating the SOPs. SOPs are a plumb line to separate safe actions from unsafe actions.
- When complex or unfamiliar decisions are to be made, a decision model (e.g. FORDEC) should be comprehensibly used by all flight crew members.
- If, because of a lack of experience or preparation, a flight crew member desires to use a decision model, this should be agreed to.
- It is difficult to revise a decision once it has been made. It becomes extremely difficult to do so during an approach following the "Continue" callout.
- Young and inexperienced pilots must enter into complex decision models more frequently than experienced pilots.
- Individual flight crew members must be familiar with "hazardous attitudes" anti-authoritarianism, impulsivity, faith in one's own invulnerability, self-overestimation, resignation, complacency and exaggerated consideration. They need to watch out for them and use counteractive measures to oppose them.
- The captain must avoid making individual decisions, actively incorporate the other crew members and always lay out his decisions for timely disposition.
- If the captain does not adequately consider the recommendations of the other crew members, they should demand due consideration.
- The crew can consciously influence the degree of harmony. Too great a degree of harmony impedes the expression of diverging opinions.
- When complex decisions are to be made, the captain should temporarily surrender his role as Pilot flying, as necessary.
- The hierarchical gradient within the crew should be actively controlled. Especially the captain must consciously be able to refrain from exercising his dominance.
- When working through the individual steps of a decision model (e.g. FORDEC), the captain should wait on presenting his view of the situation until the other flight crew members have first expressed theirs.
- The situational awareness of the crew is not the sum of that of the individual crew members, but it is limited by that of the captain's. It is essential for the other crew members to be mindful of this.
- "Double thinking" should be minimized.
- Tired crews take longer to make decisions and are more prepared to take risks than rested crews.
- The required communication within the group is heavily curtailed when conflicts remain unresolved or the level of distress becomes too great.
- Distress should be preventively avoided. If it can't be avoided, however, one's own, as well as that of the colleagues should be identified and managed.
- An error chain can be interrupted immediately by discussing the first error upon its occurrence where possible.

6.8 Empirical Experience and Trainability

Experience has been building since the 1980s through related training programs in the field of "Judgement and decision finding" (Lauber and Foushee 1981; Conolly et al. 1989).

A significant reduction in decision-related errors of up to 50 % has been realized through selection and training (FAA 1991).

A model for rational decision making (DECIDE by Lawton) has also been positively tested with respect to its effectivity (Jensen and Biegelski 1989).

On the basis of the momentary state of knowledge, we recommend the implementation of a five-tiered training program:

- The establishment of required theoretical learning objectives should be configured in written format, for instance within the framework of a flight operations handbook for line operations and as a manual for flight schools.
- The relaying and testing of theoretical knowledge within the context of the license acquisition.
- A compulsory Crew Resource Management (CRM) seminar for all license holders at intervals coupled to the respective occupational history, such as with every promotion.
- Their usage and training in flight operations with all check events and simulator refreshers. Video feedback with an appropriate debriefing is beneficial in simulator sessions.
- Consistent use of the decision model along with the awareness of possible disruptions in day-to-day line operations.

References

Boeing Commercial Airplane Group (1993) Accident Prevention Strategies 1982–1991, Seattle

Bohlman L (1979) Aircraft accidents and the theory of the situation, Resource Management on the Flight Deck, (Proceedings of a NASA/Industry Workshop, NASA Conference Proceedings 2120)

Ciompi L (2002) Gefühle, Affekte, Affektlogik, Picus, Vienna

Conolly TJ, Blackwell BB, Lester LF (1989) A simulator-based approach to training in aeronautical decision making. Aviat Space Environ Med 60(1):50–52

Damasio A (1994) Descartes' Error. Putnam, New York

Diskell JE, Sallas E (1991) Group decision making under stress. J Appl Psychol 76(3):473–478

Dörner D (1996) The logic of failure, Recognizing and avoiding error in complex situations, Metropolitan, New York

Eißfeldt H, Göters KM, Hörmann HJ, Maschke P, Schiewe A (1994) Effektives Arbeiten im Team. DLR Mitteilung, pp. 94–09, Hamburg

Elger C, Schwarz F (2009) Neurofinance. Haufe, Munich

Federal Aviation Administration (FAA) (1991) AC No: 60–22, Washington

FAA (2008) Pilot's Handbook of Aeronautical Knowledge, FAA-H-8083-25A, Oklahoma City

Flin R, O'Connor P, Crichton M (2008) Safety at the Sharp End. Ashgate, Aldershot

Gershon M (2001) Der kluge Bauch, Die Entdeckung des zweiten Gehirns. Goldmann, Munich

Gigenrenzer G (2007) Gut feelings, The intelligence of the unconscious, Penguin, New York

Hanf CH (1986) Entscheidungslehre. Oldenbourg, Munich

Jensen R, Biegelski C (1989) Cockpit Resource Management. In: Jensen R (ed) Aviation Psychology. Gower, Aldershot

Klein G (2003) Natürliche Entscheidungsprozesse. Junfermann, Paderborn

Koechlin (2011) Lecture at the DGAC Congress, November 2011, Paris

Landua R (2009) Am Rand der Dimensionen: Gespräche über die Physik am CERN, Suhrkamp, Frankfurt a.M

Lauber JK, Foushee HC (1981) Guidelines for line-oriented flight training (Vol. 1) (NASA Conference Publication 2184). Moffett Field, CA: NASA Ames Research Center

Lufthansa (1993) CF Info 1/93, Dept. FRA CG, Frankfurt a.M

Mosier K (1991) Sixth International Symposium on Aviation Psychology, Ohio

National Transportation and Safety Board (1994) Safety Study NTSB/SS-94/01, Washington

Petrilli R, Roach G, Dawson D, Thomas M (2006) The effects of fatigue on the operational performance of flight crew in a B747-400 simulator, Centre for sleep research, University of South Australia, Adelaide

Schwartz D (1989) Training for Situational Awareness. Lecture presented to the Fifth International Symposium on Aviation Psychology, Ohio State University

Telfer R (1989) Pilot Decision Making and Judgement. In: Jensen R (ed) Aviation Psychology. Gower, USA

Transportation Safety Board of Canada (TSB) (2007) Accident Report A05H0002, Air France Airbus A340 on 02.08.2005 in Toronto

Leadership and Team Behaviour

7

H.-J. Ebermann and J. Scheiderer

7.1 Introduction

"Mr. McBroom was a tyrannical boss who continuously intimidated his staff with his arbitrary moodiness. This might not have attracted attention if he had worked in an office or a factory, but Mr. McBroom was an airline captain. His co-pilots feared his temper so much that they wouldn't say a word, not even when it was obvious that the catastrophe was approaching. Ten people died when the plane crashed" (Goleman 1995).

The lone warrior who accomplishes his tasks alone and is isolated from his surroundings should be relegated to history, especially in the piloting profession. Today, we know that it is not through energy-debilitating confrontation, but only through constructive cooperation that it will be possible to fulfil our most important task: to fly the aircraft as safely as possible.

The captain is in command of the aircraft. The authority conferred upon him is accompanied by the obligation to act in a competent and responsible manner.

His closest colleagues are the other members of the cockpit and cabin crews. Together, they form a team whose concerted effort is required in order to achieve that desired result of the safe flight. There is no one member of the team who is able to achieve this all on his own.

The team is formed from a group of people who may not yet have worked with each other. The captain plays the decisive role during the process of building the team.

H.-J. Ebermann (✉) · J. Scheiderer
Main Airport Center, Vereinigung Cockpit e. V, Unterschweinstiege 10,
60549 Frankfurt, Germany
e-mail: vc.ebermann@onlinehome.de

J. Scheiderer
e-mail: scheiderer@live.de

H.-J. Ebermann and J. Scheiderer (eds.), *Human Factors on the Flight Deck*,
DOI: 10.1007/978-3-642-31733-0_7,

Over and above the aircraft, there is the "big team": This comprises an entire scope of personnel from the managerial, maintenance and fleet departments, ATC, the airport and all other co-workers required in this profession to successfully operate the aircraft.

It is in this context in the following discussion that we address the necessary requirements for an optimal "team sport" as seen from our perspective as pilots:

What is tangibly and verifiably safety-relevant and, for this reason, required of the practitioner?

7.2 Basic Principles

7.2.1 Leadership

The term "Leadership" is not clear and is oftentimes used in different ways. In literature, there are numerous definitions related to the subject of leadership. Various aspects are frequently referenced thereby, such as "the leader, the task, the situation and/or the persons being led" (Badke-Schaub, Lorei 2003). In general, leadership can be defined as the attempt to "exercise influence in order to motivate the group members to accomplish a task and, in so doing, to realize the goals of the group or the organisation"(Weinert 2004).

To lead means "to influence".

The term "to influence" generally carries with it negative connotations. It is quickly assumed that a person who can be influenced must have a weak character or a lack of self-confidence. But that's not the issue here. To influence in this context does not mean to manipulate, to sell one's colleagues on some decision or to gull them over. Neither does it mean "laissez-faire"; to leave everything as it is.

To influence means to fulfil the two fundamental, all-encompassing tasks of leadership: *cohesion* and *locomotion.*

Cohesion is the effecting and maintaining of the group association. Cohesiveness is "a measure of the strength of the members' desire to remain a part of the group"(Weinert 2004).

People don't exist merely in performance categories, but they are social beings, as well. A leader must not only *advance* a task, but he must also *get through* to his colleagues with that task. He must be socially competent, meaning he is able to cope with his own feelings, as well as with the feelings of others.

Locomotion stands for the substantive aspects of leadership. Goals need to be achieved, performance effected and services provided. It has to do with methodical competence.

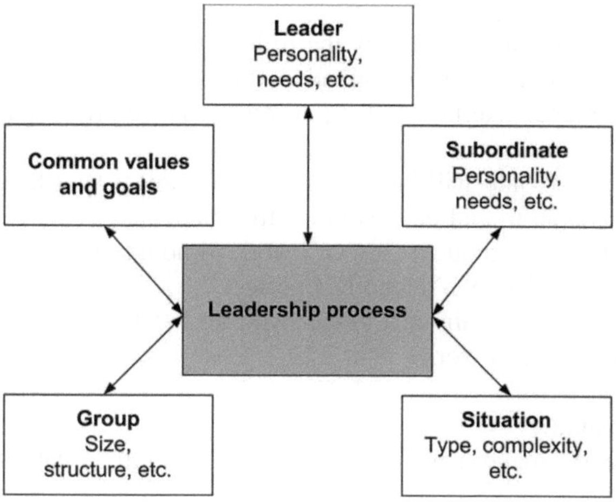

Fig. 7.1 The leadership process

7.2.2 The Leadership Process

The leadership process "incorporates reciprocal and multifaceted interaction" (Weinert 2004). The factors involved in the leadership process are depicted in Fig. 7.1.

The leadership process is understood to be the chronological sequence of the purposeful influencing of the behaviour of personnel by a leader in order to achieve a certain goal. It comprises the phases of goal-setting, planning, execution, monitoring and control.

7.2.3 The Leader

The leader, or the captain in this context, guides and influences the crew towards a common goal.

Among many other factors, his success as a leader will depend on the captain's attitude towards his colleagues. All behaviour is determined in part by attitudes. These attitudes are based on the values possessed by the captain.

McGregor was of the opinion that the style of leadership and the behaviour of the leader would depend on the leader's concept of his working people. In the 1960s, he postulated two theories about working people which encompass both convictions and attitudes: the X-Theory and the Y-Theory (Kichler 2008).

The X-Theory

The X-Theory is characterized by the following three attitudes:

- The person is naturally lazy, without initiative or ambition.
- The person tries to avoid work and responsibility wherever he can.
- The motto of the X-attitude is: Trust is good but control is better. Or: "I'm OK, you're not OK."

- **The Y-Theory**
  The Y-Theory is characterized by the following assumptions:
- The person is innovative and imaginative, if he would only be permitted to be so.
- Physical and mental exertion on the job comes as naturally to him as with sports and games.
- The person is not only willing to accept responsibility, but he seeks it out.
- The person is able to and desires to control himself.
- The motto of the Y-attitude is: It won't work without trust, self-control is better. Or: "I'm OK, you're OK."Nobody's perfect."
  A leader such as a captain or a trainer with an X-attitude would typically ask:
- What's bad about this work?
- What can I criticize?
- Who is to blame?

He'll frequently make a mountain out of a molehill; a catastrophe out of a minor mistake.

If he possesses a Y-attitude he may ask himself:

- What's good about this work?
- What can be done even better?
- How can it be done even better?

He is able to give praise and doesn't put his initial emphasis on the negative. Inevitable imperfections are used as opportunities for improvement.

Leadership functions only when one takes a general interest in his fellow human beings—without specific regard for their potential utility—and receives them with a positive underlying attitude. Because of the emphasis placed on this basic attitude, one should not strive for a leadership role in principle without such an underlying "affinity for people"(Wottawa and Gluminski 1995).

The colleagues of a leader with an X-attitude will quickly feel as if they've been attacked because of their mistakes and will respond with stress or irritation. This will soon result in their making even more mistakes. These mistakes will further confirm the feelings of the leader with his X-attitude. He will exercise even more control and impose his will to the point that his colleagues will draw back even further or cramp up. A vicious circle develops that is clearly not desirable in a commercial aircraft.

Social scientists confirm that the Y-attitude is the more effective of the two. Even with regard to the formation and retention of the group, the Y-attitude reveals itself to be significantly more effective.

A captain who finds himself to this point tending rather towards the X-attitude camp should be asking himself how he can develop more pronounced Y-attitude attributes. In order to do this, he must examine his own behaviour:

A leader who continuously nags at his staff because of small errors oftentimes has difficulty recognizing true achievement. He should consciously take the opportunity to overlook minor mistakes and give explicit praise for good performance. His colleagues will respond with an enhanced enthusiasm for their jobs and a readiness to help others. The leader's somewhat negative attitude towards his colleagues will then begin to be put into perspective (Ströbe 2006).

Getting back to cohesion:

Particularly in the flying profession, there exists a strong compulsion for rationality, as well as towards technical and logical thinking. We learn a lot about technical competency during initial and recurrent training, but very little about social competency.

The captain can expect a higher level of trust with a greater degree of respect and better communication when he

- makes an effort to understand his crew,
- takes an interest in the expectations of the individual crew members,
- considers his colleagues when making his decisions,
- establishes direct contact,
- shows himself to be sensitive to the interrelationships within the crew and
- uses criticism and feedback from the crew as a valuable instrument for monitoring himself.

7.2.4 The Colleagues

Colleagues, or crew members in this context, are the leader's partners. The same applies to crew members as for captains with regard to the areas of cohesion and locomotion. Each individual bears a responsibility for achieving the common goals with the help of technical and social competence. He independently furthers his qualifications. The "ideal" crew member works for the most part independently or autonomously. He expects the captain's support and even demands it. He is in a position to criticize his boss just as his boss is in a position to criticize him.

The working relationship between the captain and co-pilot should be particularly delineated from that of the cabin crew. The co-pilot is the captain's representative and closest staff member. In as much as the captain is responsible for demonstrating good "*leadership*", the co-pilot is responsible for demonstrating good "*followership*". This is understood to mean an "active and continuous contribution to the overall team performance, the monitoring of all changes in the system sequence and the assertion of important insights. Followership is commensurate to redundancy, permitting neither an attitude of denial nor arrogance, nor slavish submission" (Steininger 2003).

7.2.5 The Group

Commercial aircraft operations can function only as a team effort, while execution of the overall flight is a typical example of teamwork.

> "… a flying mission is always a team task" (Foushee 1982; Hackman 1987).

Literature defines a group as the pooling together of people that exhibit the following defining constituents (Rosenstiel 2007):

- Multiple persons in
- direct contact over a
- longer period of time with
- differing roles and
- common standards, values and goals; bonding together through
- a "We" feeling

A group can be comprised of:

- People that must work together, even though they may not necessarily want to.
- People who may like or dislike each other.
- People with different capabilities and characteristics.

The team extends beyond the concept of the general group. It is a group, as well, yet it is generally attributed to being a particularly well-rehearsed group (Rosenstiel 2007).

As opposed to the "normal daily office routine" where the co-workers within a group have known each other for some time, as a rule, the crew members of a flight event have oftentimes never worked with each other before. Consequently, the process of team-building is a continual one. In this context, the captain assumes the decisive role as the highest ranking leader on board. He is the point of crystallisation for the entire group. Instructions are issued by him, information is collected by him and that information is ultimately transferred on by him. He bears the largest share of responsibility for the overall success of the flight.

Through his function as captain, he assumes the lead role in the team-building process during the briefing. His behaviour will determine whether a congenial or a tense atmosphere will dominate within the crew. A particular "direction" can be established already during the greeting and with the first words uttered. It will become quickly clear whether the captain has a relaxed style of leadership or rather stern one. It is also important that he makes clear, just where he will draw the line with regard to mutual technical collaboration. In the U.S. airline industry, it is said that: "The captain sets the tone".

Highly productive groups can be recognized by the following indicators:

- A clear distribution of tasks, all group members are technically prepared, they have a common goal.
- The individual group members agree to cooperate and not to compete within the group.
- Nobody is focused on individual achievement for reasons of personal interest or ego. All group members are mutually recognized as being on an equal footing.
- There is no soliloquising, nobody is isolated. Annoying incidental remarks are not made. Discussions are on a pertinent and factual level.
- Team members bear neither open nor concealed hostilities against one another.
- Close contact is maintained to other organizational units.
- A trustful, open and free exchange of information takes place instead of a form of political secretiveness. Positive thinking is encouraged within the group.
- The leader and his prestige do not stand at the forefront, but rather the fulfilment of the task, to which end all work together to achieve.

Just how important the social climate can be within a crew is strikingly revealed in the study regarding Lufthansa's flight operations-related incidents (Lufthansa 1999) already alluded to.

7.2.6 Common Values and Goals

Corporate management and the individual captains, as well, set forth the goals and values upon which the work of the group is oriented. This should be done in a clear manner.

7.2.7 The Situation

The leader, the colleagues, the group and the goal all relate to the respective situation. All five influencing factors are reciprocally related to one another. Successful leaders adapt their leadership behaviour to suit the demands of a certain situation. This is referred to as "situational leadership behaviour".

The situational leadership behaviour model is based on the interaction between goal-oriented and relationship-oriented leadership behaviour in a specific situation.

The situation, in turn, is influenced by

- the "degree of maturity" of the colleagues/the group,
- the respective target objective,
- the organizational structure,
- the social environment.

7.3 Disruptions in the Leadership Process

7.3.1 The Optimal Working Atmosphere

A study by Lufthansa revealed that 80 % of all near-accidents (incidents) in which human error played a role were ultimately resolved through an optimal cockpit atmosphere (Lufthansa 1999).

What is an optimal cockpit atmosphere?

Within the working atmosphere of a cockpit crew, "optimal" means to have the maximum possible redundancy at all times. The wrong mixture of either antipathy or affection can lead to a constricted two-way monitoring function: One will lend inadequate support—consciously or unconsciously—to an "unappealing" colleague under certain circumstances, while a "friend" will be afforded too much trust under certain circumstances and necessary criticism will go unspoken (see illustration).

61 % of all incidents with deficient CRM are triggered by antipathy or distance. At the same time, incidents in which too great an element of trust played a role occupied 15 % of the statistics. In addition, another 26 % of the CRM problems were caused by an atmosphere considered to be too relaxed. Particularly when

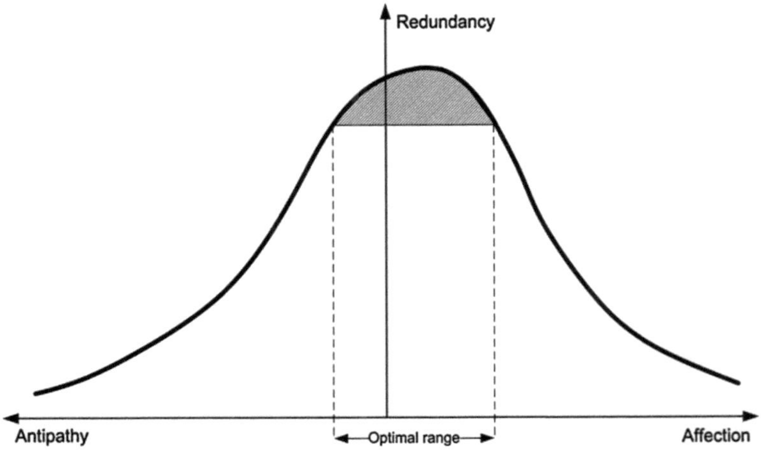

Fig. 7.2 Optimal working atmosphere

the relationship is good, it is difficult to interrupt a stimulating conversation "merely" because of an approaching airport or a descent below 10,000 ft.

An optimal working atmosphere (see Fig. 7.2) should be characterized by mutual recognition and good will. Both antipathy and resentment, as well as too great a degree of familiarity and "camaraderie" should be avoided.

The risk of a safety-relevant incident increases by a factor of 5 in a suboptimal working atmosphere!

The study recommends that, in the context of CRM seminar training, each pilot should develop a personal mental process that he can resort to with respect to a situation. Moreover, social competency should receive greater emphasis during the pilot selection process.

Furthermore, if redundancy is constrained because of too great a degree of affection (e.g. when nearing the end of cruise flight just prior to initiating the approach), VC recommends a reduction of private communication and a strict adherence to the Sterile Cockpit concept.

It becomes more difficult when redundancy is constrained due to antipathy, which can even arise on a temporary basis. Small talk in an appropriate manner should be pursued at first in order (re)establish a common foundation. If this is not successful, a disruption in the working relationship should be openly addressed. Just how to do this in the most effective and diplomatic manner can be learned in a VC training seminar, for example. If this, too, fails to achieve the desired results, then the only alternative is to disband the team as quickly as possible.

7.3.2 Conflicts

Conflicts are an integral part of the leadership process. They occur within the individual and within the group, as well as between goals and values. They are a universal, necessary element within the social structure (Dahrendorf 1962).

Conflicts comprise two fundamental aspects:

- They are a prerequisite to change. They release energy and activity. They release ideas. They have a "purifying" effect.
- On the other hand, in extreme cases they can lead to instability and stress, thereby encouraging inappropriate emotional behaviour (Rosenstiel 2007).

For this reason, the attitudes of the people involved play a very important role in the search for fruitful conflict resolutions. People who get along well with one another, who respect each other and who enjoy working together are more likely to pursue the attainment of a common goal than those who are indifferent to one another.

Unfortunately, conflicts are often accompanied by negative feelings such as anxiety, displeasure, irritation and uneasiness. This is all too human, but it also harbours the risk of uncontrolled defence reactions arising within the conflict partners. These can be the urge to:

- **Fight**

  The opponent is to be mentally (under some circumstances perhaps even physically) destroyed. The instruments used for this are crosstalk, defamation or false information (and, under some circumstances, even violence).
- **Flee**

  The conflict is seemingly resolved through evasion. Conflicts are avoided; each person stares out his own window and no social communication takes place.
- **Resign**

  Conflicts are suppressed, played down, compensated for by concentrating on other things or just plain given up (mental resignation). These defence reactions have one thing in common: they may divert the individual's attention away from his negative feelings, but they do not address the fundamental problem; they impede any benefit from being gained out of the conflict.

It cannot be the objective to emerge out of a conflict as the winner, because the win-lose mentality destroys relationships and, in so doing, permanently inhibits optimal productivity.

Therefore, it is extremely important to actively seek resolution to a conflict in the cockpit rather than evading it.

Two assumptions must be understood in order to productively deal with conflicts:

- Conflicts do not necessarily arise from generally irreconcilable points of view. This means that conflict adversaries are not necessarily conflict opponents, but rather partners in a phase that is characteristic of a normal working relationship.
- It is usually not the actual needs and goals that clash during conflicts, but rather the resolutions already attempted with respect to those needs and goals. Thus, the conflict partners must learn that there are several methods for feasibly resolving each need and goal.

Ideally, conflicts should be approached rationally with a bilateral resolution being pursued so that the end result will ultimately be a beneficial one.

Step 1: Exchange

The bilateral approach referred to above means an exchange of opposing opinions, without resorting to valuations, lecturing or conversions in the process. Attention should first be drawn to the things shared in common before disruptive issues are debated. The individual's own interests and motives should be presented in a comprehensible manner. It is possible to promote one's own interests while showing understanding for the motives and messages of the other, yet without expressing consent in so doing. "Understanding the viewpoint of the other side does not mean consenting to it" (Becker 1995). "I-messages" and "active listening" can be very helpful in this regard. The presentation of thoughts, situations and feelings in the I-form helps the partner to comprehend the situation as the other person is experiencing it.

Active listening, as a counterpart to the I-message, means to provide the other person with feedback from his messages in order to eliminate misunderstanding.

This will help to create a common foundation.

In conflicts just as with communication, there will always be a factual level as well as a relational level. At the same time, the relational level is actually afforded overriding significance above the factual, objective level in the unconscious mind (Watzlawick 1983)! For this reason, the I-messages and feedback should be formulated as precisely as possible.

Step 2: Seeking resolution

In the process of seeking possible resolutions satisfactory to both sides, it must be borne in mind that mutual recriminations, concealed or open criticism and the search for someone to blame are counterproductive. They immediately re-spark negative feelings and, in so doing, produce defence reactions.

A positive path towards conflict resolution is based on

- An underlying attitude within the conflict partners that interpersonal relationships are an essential element of productive cooperation.
- Open relationships. Intense and irreconcilable experiences from the past put a burden on and complicate productive bilateral solutions.
- The preparedness to seek solutions, even if these are not immediately apparent.

The model for realizing positive benefits from the unavoidable conflicts experienced within the leadership process draws in both partners. It is absolutely essential that this be conveyed during the initial training provided by an air carrier so that all company employees are afforded the same insights with regard to conflict resolution.

One more aspect: Once a team has experienced success in resolving one conflict, it will be much easier for them to handle subsequent conflicts.

7.3.3 Groupthink

Just as a well developed sense of belonging can stimulate a crew to extraordinary achievement, the phenomenon of peer pressure can have just the opposite effect and greatly obstruct team performance. The concept of groupthink was developed by Janis in 1972 based on analyses of incorrect political decisions. The main proposition is that "groups tend towards error-prone decisions when the feeling of togetherness within the group becomes more important than critical scrutiny" (Badke-Schaub, Hofinger, Lauche Badke-Schaub et al. 2010).

Accordingly, groupthink is present when the members of a cohesive group have formed an opinion—no matter how false or unrealistic—which is maintained even in the presence of convincing arguments against it. In so doing, the members adapt so readily to the peer pressure that they are no longer able to think critically (Weinert 2004).

The reason for this is that the individual behaves differently as soon as he is no longer alone, but enters into social relationships. He looks for bonds and forms coalitions in order to better cope with his own insecurities and to find affirmation. "The striving for unity suppresses any motivation to consider other possible alternatives" (Weinert 2004).

This conceals risks—especially in safety–critical areas (on the basis of Badke-Schaub, Hofinger, Lauche 2008):

- Not all possible solutions are discussed. "Poor and erroneous decisions are the result" (Weinert 2004).
- Inadequate consideration and incomplete review of action alternatives.
- Selective information processing and deficient information research. Only information that substantiates an opinion will be incorporated into the decision.
- Expert information to supplement the individual's own information is not sought after.
- The reassessment of previously discarded alternatives is lacking. A course already adopted is considered to be irreversible and is not subjected to ongoing reappraisal.

Eight symptoms of groupthink exist (according to Weinert 2004):

- The group feels itself to be invulnerable and, for this reason, tends toward optimism and an increased willingness to take risks.
- Rationalisations are construed in order to devalue other sources of negative information.
- Faith is placed in the morale of the group. Opposing views are considered to be evil. The moral and ethical consequences of one's own actions are ignored.
- Stereotyped perceptions are formed. Other groups are portrayed as being of bad character, thereby complicating reasonable negotiations between different groups.
- There is moral (peer) pressure on the minority. Well-founded criticism from dissenting persons is suppressed.
- Minorities sensor themselves. They play down inner, by all means justifiable, doubts.
- An illusion of unanimity prevails.
- "Thought guardians" exercise strong influence by virtually firewalling any information that would disturb the self-satisfaction of the group.

Measures that oppose groupthink

- Each group member is a "critic" who should immediately articulate any doubts.
- The leader consciously reserves voicing his own opinion at first in order to avoid influencing the other members.
- Heed is given to the symptoms of peer pressure and they are done away with immediately.

7.3.4 Normative Influence

Normative influence is related to groupthink in that it represents the influence of the majority or the more powerful entity. An opinion held by the majority or the more powerful entity that diverges from one's own convictions sets in motion a process of comparison within the person being influenced based on the striving for a positive self-worth. The result is oftentimes the uncritical assumption of the alien opinion.

This assumption is not based on conviction, but on identification with the majority or with the more powerful entity, and on the positive self-worth that comes along with it. Attention is directed to the person who voiced the opinion, while its content is of only peripheral importance. For this reason, opinions held by supervisors are more readily accepted than those held by his colleagues. **Even when they are discernibly wrong!**

The dangerous effects of this mechanism are systematically made use of in dictatorships through their propaganda.

In the flying profession, the most common individual mistake made by FOs in conjunction with total loss accidents was their failure to critically evaluate the captains' decisions (NTSB 1994).

This is understandable when considering the impact the normative influence can have due to the elevated status of the captain. The FO will oftentimes suppress necessary and warranted criticism in his subconscious drive to identify with the captain.

It is precisely during the time-critical phases of flight, such as during a hectic takeoff or an unstabilized approach, that only limited time is available for critical deliberation. The consequence is that the FO retreats too rapidly back into this unconscious action pattern.

It becomes particularly hazardous when the captain demonstrates a casual attitude towards dealing with safety regulations. The resulting rate of error increases significantly, as numerous studies reveal.

Albeit, having a casual attitude towards dealing with safety regulations is commonly seen as being a sign of particular expertise—a fatal error.

For this reason, experience and sovereignty are oftentimes linked together unconsciously (Helfrich-Hölter 1996).

It must be clear to every captain that a lax attitude in dealing with Standard Operating Procedures is in no way a sign of expertise or worthy of admiration, but it attests to a general degree of carelessness. The consequence for the professional captain is that he has no other choice but to abide by the SOPs with great consistency.

What solutions are available?

An opinion asserted in opposition to the normal hierarchy, meaning "from the bottom up", such as from the FO to the captain, may certainly be taken into account, but only if it has been clearly expressed and pervasively represented.

What can a captain do?

It must be clear to every captain that his colleagues in the cockpit are subject to the dilemma of normative influences. Therefore, the captain must create a working climate that encourages his crew to provide criticism whenever they deem it appropriate. He should actively invite criticism already during the briefing. This makes it possible for the captain to give a clear signal early on and to manifest his openness. A briefing patterned after "we do everything according to standards" does not do this adequately.

Furthermore, the captain should encourage criticism by openly disclosing those factors that lower his threshold to stress and overload, by addressing his own mistakes and offering a "thank you" for every objection or callout made.

What can co-pilots, in particular, do?

A dissenting opinion should be clearly expressed and pervasively represented. Strict adherence to the SOPs with their corresponding callouts is essential to this end.

7.4 Leadership Style

Successful and efficient leadership will always be based on a certain style of leadership. This is understood to be "a typified and recurring leadership behaviour exercised within bandwidths and leadership contexts" (Wunderer 2007).

Efficiency is understood to mean the degree to which the leader achieves the agreed upon results. A leader is not efficient when he just appears to be efficient or when he desires to achieve only his own personal goals. Efficiency is much more the result of an accurate determination of a specific leadership situation and the corresponding influence exercised.

Every employee and every leader can be distinguished by their respective behaviour. The individual style of leadership responds to these behavioural distinctions.

The bandwidth of leadership styles represents the capacity for modifying leadership behaviour. A large bandwidth is desirable. It enables the leader to cope with different situations and, in so doing, to lead more efficiently. The following discussion provides a brief introduction to the most well known dimensional approaches and their leadership styles.

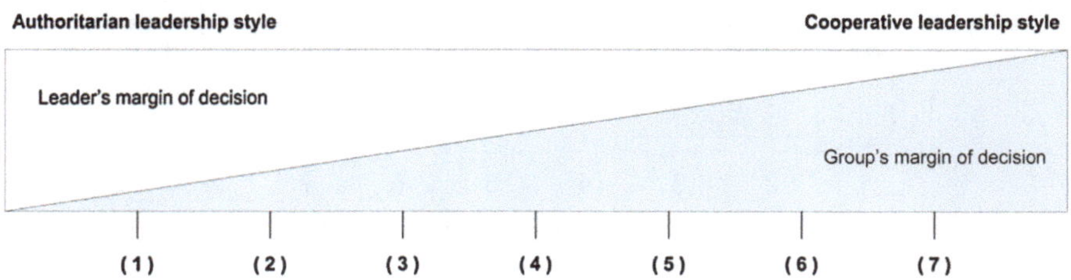

Fig. 7.3 One-dimensional leadership style

7.4.1 One-Dimensional Leadership Style

A one-dimensional leadership style (see Fig. 7.3) distinguishes merely according to the degree of authority (Wunderer 2007).

1. The supervisor exercises an authoritarian style.
2. The supervisor makes the decision with understated authority.
3. The supervisor makes the decision but asks his staff members for their opinions.
4. The supervisor makes an interim decision, but is open for changes.
5. The supervisor asks for recommendations from his staff members, but then makes the decision on his own.
6. The supervisor establishes the decision-making margin, within which the group decides.
7. The group makes the decision within the margin determined by the higher authority (higher-level supervisor).

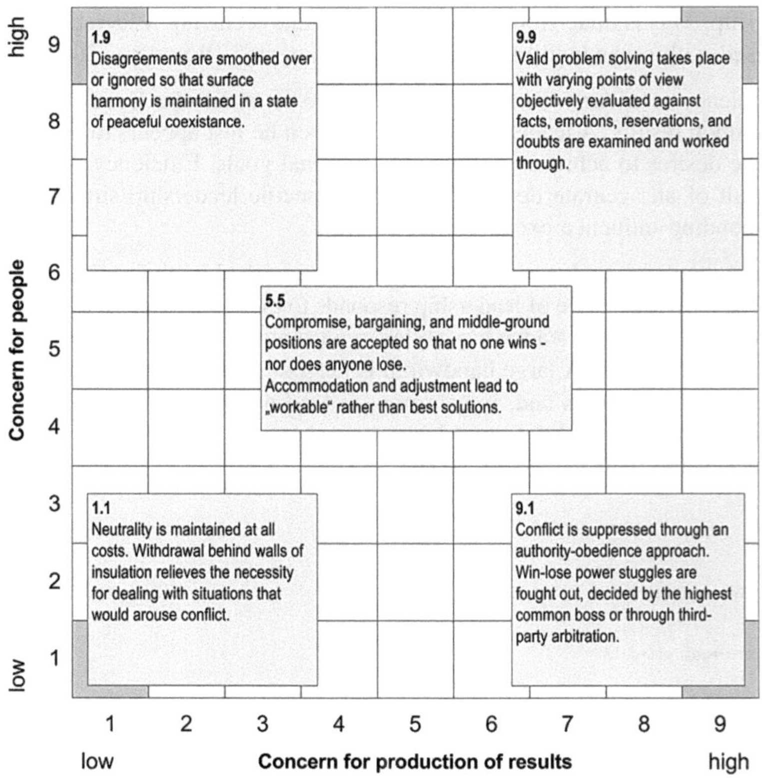

Fig. 7.4 Managerial grid

7.4.2 Two-Dimensional Leadership Style

The so-called two-dimensional approach proposed by Blake and Mouton in 1980 (Weinert 2004) considers two behavioural dimensions. They are portrayed in a behavioural grid (*Managerial grid*), whereby the vertical axis represents person-oriented leadership behaviour, while the horizontal axis represents task-oriented leadership behaviour. This depiction results in five style-extremes (see Fig. 7.4).

The **maturity-related conception** represents another two-dimensional approach according to Hersey and Blanchard (1977). It is directly linked to the concept of the *Managerial grid* and adds the maturity of an employee in a specific situation as another factor to be considered (see Fig. 7.5). In so doing, a differentiation is made between four levels of maturity (Göters et al. 1994), whereby the level of maturity is understood to mean "the ability and willingness of those being led to assume responsibility for controlling their own behaviour" (Weinert 2004). "It is the present state of existing competence and of identification with the task" (Weinert 2004).

- Low level of maturity (motivation, knowledge and ability are lacking).
- Low to moderate level of maturity (motivation, but a lack of ability).
- Moderate to high level of maturity (ability, but a lack of motivation).
- High level of maturity (motivation, knowledge and ability all exist).

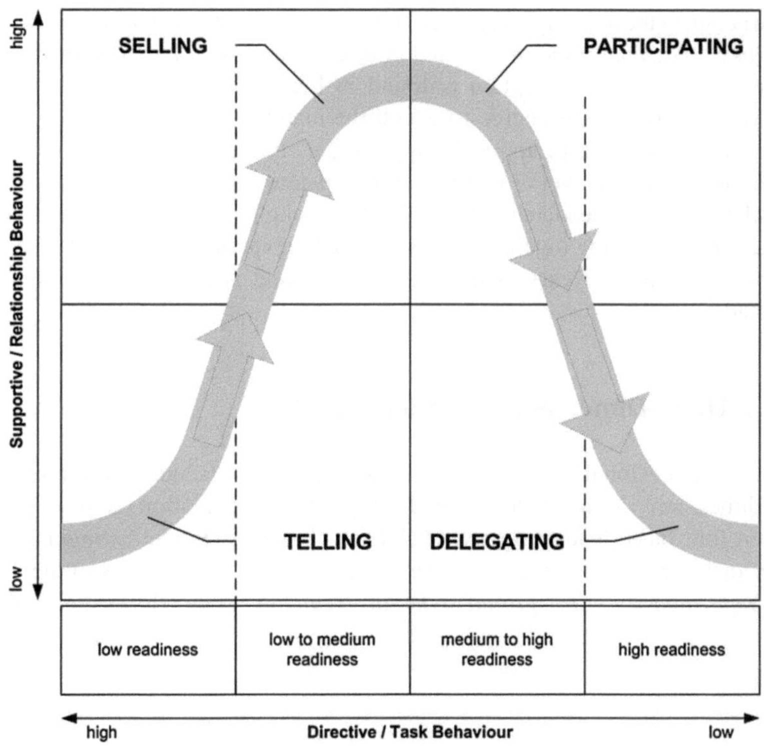

Fig. 7.5 Level of maturity model

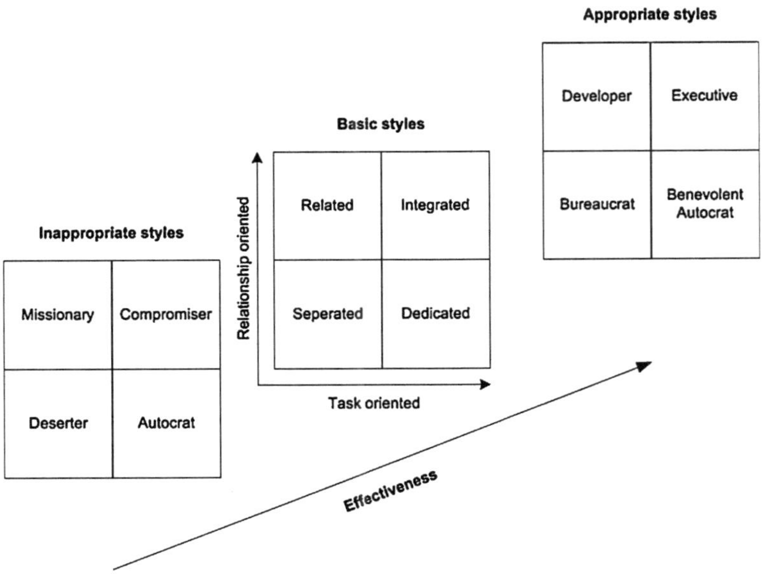

Fig. 7.6 3D Theory conception

The height of the maturity level is determined by three criteria: *motivation to achieve, willingness to assume responsibility* and *training and/or experience* (Weinert 2004).

The "artistry" of the captain is found in his ability to correctly estimate or intuitively determine the degree to which the maturity or willingness of the crew has developed in order to apply the appropriate style of leadership.

The inclusion of a level of maturity or competence becomes an important issue, particularly for the leadership tasks in the cockpit. It makes a big difference whether one is working together with a new and inexperienced co-pilot or with one who possesses many years of flight experience and is perhaps just about to be promoted.

7.4.3 Three-Dimensional Leadership Style

The three-dimensional approach (Reddin Reddin 1981) includes **effectiveness** as a third dimension in the style of leadership. A differentiation is made thereby between four basic leadership styles. All four basic styles are distinguished by a greater or lesser degree of effectiveness (see Fig. 7.6). This results in the *Related style*, *Separated style*, *Integrated style* and *Dedicated style* (Rosenstiel 2007).

The **Related style** places value on good interpersonal relationships.	
Effective	Ineffective
The encourager is not self-serving in his view towards colleague development and delegates to the greatest extent possible.	The people pleaser neglects his colleagues.
The **Separated style** puts confidence first and foremost in the process	
Effective	Ineffective
The administrator epitomizes the bureaucrat.	The wimp relies on compliance with rules and regulations.
The **Integrated style** strives for a balance between person-orientation and task-orientation.	
Effective	Ineffective
The integrator motivates and encourages his colleagues.	The compromiser shies away from decision-making and avoids confrontation.
The **Dedicated style** places performance-orientation at the forefront.	
Effective	Ineffective
The man of action is persuasive because of his expert knowledge.	The autocrat overwhelms his colleagues.

7.5 Safety-Relevant Behaviour in the Day-to-Day Working Environment

The references made to most of the behaviour patterns in this section are the result of accident analyses. If the sources are not separately referenced in the footnotes, then they are taken from a study conducted by Prof. Helmreich of the NASA/University of Texas (1995). Specific jet aircraft accidents are listed therein, upon which these behaviour patterns are based and which could have prevented had they been applied.

The desirable safe behaviour is *italicized* in the sections that follow.

7.5.1 The First Impression

We unconsciously appraise a person within the first seconds of seeing him. The person's outer appearance, his behaviour and our past personal experiences with similar persons all influence us. Among other things, we develop a picture of sympathy or antipathy, seriousness or laxity.

A nonchalantly worn uniform, a superficial "good day", an over-exaggerated formality, the posture, the artificiality, speech that is too loud or too soft; these all put one quickly in a potentially false, bad light.

Pilots actively endeavour to obtain an authentic first impression (Göters 1994*).*

7.5.2 The First Conversation

During the first conversation—commonly at the initial joint flight preparation—the foundation is laid for the overall working relationship.

The desire and the will to work together in a team is built up and, later, cultivated.

Attentive and patient listening is practiced.

There are no interruptions.

An inappropriate element of haste is not being propagated.

Eye contact is maintained where expedient.

The conversation does not only take place at a factual, objective level.

7.5.3 Briefings

Briefings take place on a recurring basis within the cockpit crew and together with the entire aircraft crew prior to and during the flight. In the broad sense of the word, informative passenger announcements are even briefings.

Briefings are generally appropriate for the operation, meaning they may not be excessively formal.

They are formulated to be interesting, interactive and composed, and do not comprise frequently recurring platitudes ("we do everything according to standards", "communication is important to me", etc.).

Potential problems and ambiguities are preferably discussed and clarified in advance.

Deviations from normal operations are clarified in advance.

The crew prepares itself for special technical situations (e.g. unusual weather, particular anomalies on approach, etc.*).*

7.5.4 Standard Operating Procedures

All crew members are actively concerned about SOP compliance (JAR-TEL 1999).

Intervention is exercised in the event of an SOP deviation.

7.5.5 Actively Striving for Teamwork

Trust and dependability must be established. They do not result automatically.

The group climate is appropriate for the phase of flight. Conversation outside the critical flight phases take place in compliance with the "Sterile Cockpit" concept.

Support is called for.

Support is offered at one's own initiative.

Motivation is encouraged and praise is given.

Colleagues are immediately and readily informed where necessary (Göters 1994*).*

Questions and criticism are encouraged.

7.5.6 Advocacy

This is understood to be the appropriate and active representation of one's own standpoint. Particularly young and inexperienced colleagues will naturally have problems with this now and again. Rhetorical superiority does not necessarily mean being right.

One's own standpoint is represented even at the risk of possible conflict.

An opinion should be substantiated in such a way that the colleagues can comprehend it.

The captain actively ensures that each person's opinion is objectively appraised for its content.

Nothing is taken 'lightly"; no remarks or opinions are ignored.

Nobody is ridiculed.

The captain neither induces a state of anxiety—even unconsciously—nor does he put up with submissive behaviour on the part of his colleagues (Lufthansa 1999*).*

The captain offers thanks for criticism (and dissenting opinions) and, in so doing, motivates the next message.

Communication is open and tasks are never approached with one colleague working against another, even with a poor group climate.

7.5.7 Assertiveness

The term assertiveness is closely related to advocacy. In this context, it is understood to signify the determination behind the stand taken.

Enquiry is made with appropriate clarity and at a commensurate volume; one's own insights and opinions are expressed (Hackman 1993*).*

Here, too, the captain must expect that especially inexperienced colleagues will express themselves in an almost incomprehensible, shy and even non-verbal manner.

Correspondingly, it is essential that he take these signals seriously, enquire about them and appraise them without reservation for their objective content, independent of their perhaps less than convincing form.

7.5.8 Optimal Hierarchical Gradient

The optimal hierarchical gradient was already discussed at length in the chapter on Communication.

The captain actively adapts his dominance to the other (cockpit) colleagues.

An appropriate balance is maintained between authority and collaboration.

If the captain cannot act with sufficient dominance or maintain this balance, then the other crew members will draw back.

The captain's role as a leader is not—openly or concealed—called into question.

7.5.9 Setting Priorities

Issues of secondary importance with respect to flying, such as passenger needs, snack breaks, company communications, etc., are taken care of or given preference at the appropriate time.

The Flight Management System (FMS)—if it becomes a hindrance—will be switched off or its degree of automation reduced.

Available capacity will be used for operation of the FMS during critical flight phases only when it is required for fulfilling the task of flying the aircraft.

Problems that emerge are identified as quickly as possible (Göters 1994*).*

7.5.10 Feedback

Positive and negative feedback is provided at a convenient time and in an appropriate manner.

Feedback is specific, objective, based on observed behaviour and constructive.

Feedback is accepted willingly and not defensively.

Criticism is offered in private. Nobody is denounced in front of others (Condor Izmir accident 1988*).*

Callouts will not be taken (or given) as criticism.

7.5.11 Professional Conflicts

Conflicts will be dealt with at the objective level.

Mistakes are admitted to.

All participants actively endeavour to resolve conflicts.

7.5.12 Social Climate

Aggravation over cockpit colleagues is an alarm signal and significantly degrades the mutual redundancy (Lufthansa 1999). It is degraded, likewise, when too casual of an atmosphere reigns.

Disruptions within the group climate to the point of even personal animosities must be addressed as early as possible ("I don't feel good about our working relationship"; "I have the feeling that unspoken problems exist").

A poor working atmosphere is not accepted.

Deliberate attention is drawn towards animosity once it is recognized.

The optimal working atmosphere is actively controlled through such measures as complying with the "Sterile Cockpit" concept on the ground and below 10,000 ft, as well as by giving personal attention and sharing in small talk outside these phases.

7.5.13 Stress

Crew members should be able to recognize their own signs of stress as well as that of their colleagues.

One's own state of stress or overburden is immediately disclosed to the other colleagues ("I'm not that far along yet." "Can you take over?" "Things are happening too quickly for me right now.").

7.5.14 The Lone Warrior Syndrome

Pilots who tend to refrain from discussing or coordinating matters with the other colleagues, meaning pilots who try to do things single-handedly (usually the captain), present a problem. In practice, this is not usually the result of "ill will", but is rather practiced with the best intentions for the operation of the aircraft in time-critical situations.

Hectic and haste are not tolerated on-board the aircraft.

One's own actions are commented upon, while intentions and plans are explained (Göters 1994*).*

7.5.15 Redundancy

If something such as an ATC clearance is understood by just one pilot, then it will always be verified.

It shall be ensured that one pilot is always responsible for controlling the aircraft.

Distractions are avoided.

Simultaneous activities by all pilots (e.g. sorting of charts, checking circuit breakers, etc.*) are avoided.*

References

Badke-Schaub P, Lorei C (2003) Führung und Entscheidung, In: Strohschneider S (Pub.): Entscheiden in kritischen Situationen, Verlag für Polizeiwissenschaft, Frankfurt

Badke-Schaub P, Hofinger G, Lauche K (2010) Human factors psychologie sicheren handelns in risikobranchen. Springer, Heidelberg

Becker H, Hugo-Becker A (1995) Psychologisches Konfliktmanagement, Menschenkenntnis, Konfliktfähigkeit, Kooperation, 2. completely reworked and expanded edition, Beck Wirtschaftsberater in dtv-Verlag, Munich

BFU Bundesstelle für Flugunfalluntersuchung (German federal body for accident investigation) (1988) Investigation of the Condor accident in Izmir, Turkey on 02 Jan 1988

Blake R, Mouton J (1980) Verhaltenspsychologie im Betrieb, Duesseldorf and Vienna

Dahrendorf R (1962) Gesellschaft und Freiheit Zur soziologischen Analyse der Gegenwart. Piper, Munich

Foushee HC (1982) The role of communications, socio-psychological and personality factors in the maintenance of crew coordination. Aviat Space Environ Med 53(11):1062–1066

Göters KM, Eißfeld H, Hörmann HJ, Maschke P, Schiewe A (1994) Effektives Arbeiten im Team: Crew Resource Management for Pilots and Air Traffic Controllers, Message 94-09, Internal publication for the Deutschen Zentrum für Luft- und Raumfahrt (DLR—German Aerospace Center), Hamburg

Goleman D (1995) Emotional Intelligence. Bantam Place, New York

Hackman JR (1987) Group-Level Issues: the design and training of cockpit crews. In: Orlady, HW, Foushee HC (eds) Cockpit Ressource Management Training, NASA Conference Publishing, pp 23–39

Hackman JR (1993) Teams, leaders, and organizations: new directions for crew-oriented flight training. In: Wiener EL, Kanki BG, Helmreich HL (eds) Cockpit Ressource Management. Academic, Orlando, pp 47–69

Helfrich-Hölter H (1996) Menschliche Zuverlässigkeit aus sozialpsychologischer Sicht. In: Zeitschrift für Psychologie, 204 pp 75–96

Helmreich R et al (1995) Behavioural markers in accidents and incidents. NASA/University of Texas, Houston

Hersey P, Blanchard KH (1977) Management of organizational behaviour, Utilizing human resources, Prentice-Hall, Englewood Cliffs

Kichler E (Pub.) (2008) Arbeits- und Organisationspsychologie, 2. corrected edition, Facultas Verlags- und Buchhandels AG, Vienna

Lufthansa (1999) Cockpit Safety Survey, Department. FRA CF, a.m, Frankfurt

NTSB, National Transportation and Safety Board (1994): Safety Study NTSB/SS-94/01, Washington DC

Reddin WJ (1981) Das 3-D-Programm zur leistungssteigerung des managements, verlag moderne industrie, Landsberg am lech

Rosenstiel LV (2007) Grundlage der Organisationspsychologie, 6th edn. Schäfer-Poeschel, Stuttgart

Ströbe R (2006) Grundlagen der Führung: mit Führungsmodellen, 12th edn. Windmühle, Hamburg

Steininger K (2003) Führung und Zusammenarbeit im Flugbetrieb, Crew Resource Management für Berufs- und Verkehrsflugzeugführer nach JAR-OPS 1.934, Books on Demand, Hamburg

Watzlawick P (1983) The pursuit of unhappiness. Norton, New York

Weinert AB (2004) Organisations- und Personalpsychologie, 5. completely reworked edition, Beltz, Basel, Weinheim

Wottawa H, Gluminski I (1995) Psychologische Theorien für Unternehmen. für angewandte Psychologie, Göttingen

Wunderer R (2007) Führung und Zusammenarbeit: eine unternehmerische Führungslehre, 7. reworked edition, Luchterhand, Cologne

8 Fatigue and Alertness Management

Hans-Joachim Ebermann and Maria-Pascaline Murtha

8.1 Introduction

Fatigue was officially declared as the main cause of a commercial aircraft accident by the U.S. National Transportation Safety Board (NTSB) for the first time in 1994. Since that time, fatigue has been systematically queried as an ongoing part of the accident investigation process.

With the continual shift changes and overly long duty times associated with an industry that works around the clock, almost every pilot today is faced with the challenge of coping with fatigue, regardless of whether they fly short-, medium- or long-haul routes. Furthermore, the number of flights per year per pilot has increased significantly due to the pressures of competition.

Human beings are the only component in the aviation industry that weren't created to function around the clock.

It was against this background that the VC/DLR Alertness Management Training was introduced. This program provided the impetus to define and present the most important facts related to the subject of sleep in this chapter. The connection between sleep, fatigue and performance will then be described, after which methods will be proposed for safely coping with them.

The following sections are limited to the issues of concern to flight crews and purposely omit the consequences and requirements for the flight operations organisation (roster-design, minimum rest periods, crew hotel criteria, etc.) in order not to go beyond the scope of the discussion.

H.-J. Ebermann (✉) · M.-P. Murtha
Main Airport Center, Vereinigung Cockpit e. V,
Unterschweinstiege 10, 60549 Frankfurt, Germany
e-mail: vc.ebermann@onlinehome.de

M.-P. Murtha
e-mail: mp.murtha@yahoo.com

H.-J. Ebermann and J. Scheiderer (eds.), *Human Factors on the Flight Deck*,
DOI: 10.1007/978-3-642-31733-0_8,

8.2 Definitions

8.2.1 Internal Clock, Circadian Rhythm and Jet Lag

Our internal clock regulates many of the body's functions and is controlled from the brain's suprachiasmatic nucleus. This is where our hormonal equilibrium, our digestive system and our body temperature are synchronized.

One of the body's most important cycles is generated in the process: the circadian rhythm (from the Latin: circa = about, around, dies = day), which equates to an average of 24.25 h in humans. The most important factors influencing this rhythm are light and darkness. Other time-related factors exercising less influence on the body's function and rhythm are meals, as well as social and physical activities (Hawkins 2002).

The phase of time referred to as the circadian low point lies between 3 and 5 a.m., during which time the body is programmed for sleep. It is also during this time that a significant decrease in the performance capability is discernible in most people. For this reason, demanding tasks should not be scheduled for this period of time.

The hormone melatonin is of great importance with regard to fatigue because it controls the human body's day/night rhythm. This hormone is secreted in a recurring 24 h rhythm (a time cycle that varies somewhat depending on the person) and influences sleep.

Jet lag is closely tied to the internal clock. It is the consequence of a desynchronization in the day/night rhythm when flying through different time zones and is provoked by a disruption of the body's individually distinctive, day–and night-related, light/dark cycle. Its symptoms are physical and mental in nature, such as a curtailed, restless sleep, premature wakening, drowsiness and reduced motivation, and influence the individual's quality and quantity of sleep, which impacts fatigue in turn.

As a rule, it is easier to put off the sleeping and waking times until later, meaning the day is extended rather than curtailed. For this reason, a crew will be better able to adapt to the local time at its destination on a westbound flight than on an eastbound flight (Samel et al. 1997).

There are several factors that have a significant influence on the characteristics of the jet lag: the direction to the destination (East or West) and the physical condition of the traveller.

As a general rule of thumb: without targeted management, a person will need an average of one day for each time zone flown though in order to recover from jet lag (Strauss 2009).

Physical effects will already begin to appear on trips crossing 1–3 time zones. It is not by chance that the road traffic accident rate increases on the first day of summer daylight time (the night is curtailed by 1 h) (Rosekind et al. 1995).

8.2.2 Fatigue

"Fatigue is a condition of reduced mental or physiological working capacity arising from sleep deprivation, overly long periods of alertness and/or physical activity. It can limit a flight crew member's wakefulness and his ability to safely operate the aircraft or carry out safety-relevant tasks"(EASA).

It comprises tiredness, drowsiness, exhaustion or weariness. One reason for this could be that a person is not getting enough sleep; either qualitative or quantitative. It could also be the result of a disturbed biological rhythm (due to work times contrary to the circadian rhythm), a demanding physical or mental task, boredom or monotony.

8.2.3 Sleepiness

Sleepiness expresses itself when a person has great difficulty staying awake or shows a significant disposition for falling asleep. Sleepiness can be abated through physical activity or with caffeine, but it can be fully offset only through sleep.

Of danger in this regard is that one assesses his own abilities and performance as being far better than they are physiologically proven to be (Rosekind et al. 1995).

8.3 Sleep

Sleep is a physiological need in humans and is of vital importance for the body's physical and psychological regeneration. Animal experiments reveal that permanent sleep deprivation can result in death. Sleep does not merely consist of one

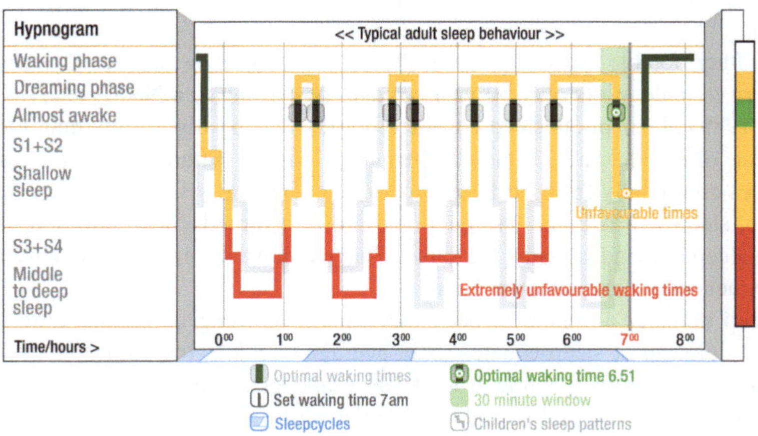

Fig. 8.1 Phases of sleep (© f-ax advertising)

phase of sleep, but it is subdivided into various stages. Depending on the scientific conviction, reference is made to 2 or 3 stages (see Fig. 8.1): the waking phase, REM (rapid eye movement) sleep and NREM (non-rapid eye movement) sleep.

8.3.1 Phases of Sleep

NREM sleep is characterized by little or slowed physical and mental activity. Breathing, the heartbeat and brain activity are all reduced. It is a phase in which the person doesn't dream and which is required for physical regeneration. NREM sleep is sub-divided into 4 levels: Level 1 is designated as the falling asleep phase, while levels 3 and 4 are the deep-sleep phases (VC/DLR Alertness Management).

If a person is awakened during phase 1, he will be able to comprehend his existing situation relatively quickly. If he is awakened during sleep phase 3 or 4, however, he will need more than 15 min to awaken from so-called sleep inertia and to recover his full performance capabilities.

REM sleep is referred to as dream sleep. It is characterized by the activation of different cerebral areas and serves the person—simply stated—in the process of mental restoration. In this state, the brain is as alert as it is while these lines are being read.

The REM and NREM sleep periods in adults alternate at about 90 min intervals. After 60 min of NREM sleep, there follows just under 30 min of REM sleep (Rosekind et al. 1995).

8.3.2 Quality and Quantity of Sleep

It is not the quantity of sleep, alone, that is of crucial importance, but the quality of sleep as well. This means that factors such as the calmness and duration of the REM and NREM phases will greatly influence our quality of sleep (Hawkins 2002).

Another notable aspect related to the quality of sleep is, above all, a matter of concern to male pilots over 50 years of age. According to a NASA study (Gander et al. 1989), these pilots sleep considerably less during flight tours than their 20–30 year old colleagues, which corresponds to the natural sleep process that changes continually from birth on into old age. Generally speaking, after 50 years of age, sleep will become less deep and more disturbed. Sleep duration declines while sleep demand remains the same. This can have severe operationally-related consequences for the affected pilots.

Sleep demand differs with each individual according to various criteria. The natural duration of sleep equates to an average of 8.5 h. Individual variations are genetically determined, while around 100 years ago it was almost 1 h longer than it is today. Our way of life and the influence of artificial light have contributed substantially to these changes.

8.3.3 Sleep Hygiene

Sleep hygiene refers to sleep-promoting lifestyle habits and behaviour patterns. An example can be relaxation activities and a light meal just before "going to bed", as well as ensuring a pleasant sleeping environment.

There are numerous strategies that can be used to encourage sleep hygiene (VC/ DLR Alertness Management).

Prior to going to bed:

- Ensure a darkened and quiet atmosphere with a comfortable room temperature.
- Avoid caffeinated beverages, alcohol and heavy meals.
- Do not go to bed hungry.
- Avoid strenuous mental and physical activities.
- Do not watch disturbing television programs.
- Do not take a "long" sleep in front of the television before going to bed.
- Sport activity—about 2–3 h before going to bed—can be helpful in promoting better sleep. After this time, however, such activity will initially impede falling asleep.
- Develop a personal bedtime ritual, such as reading a book, listening to music or the like.

8.3.4 Sleeping Disorders

There are often several reasons for a restless sleep. Such sleeping disorders can result from outside influences, such as one's personal surroundings, or they may be medically related, such as is the case with sleep apnoea and insomnia.

Sleep Apnoea

Sleep, in this context, is accompanied by breathing interruptions (apnoeas) for a duration of several seconds (sometimes even up to 90 s). The breathing interruptions lead to an oxygen deficiency resulting in an abrupt awakening. The person will have no recollection of this in the morning, however.

The structure based on the depth of sleep and the phases of sleep is disrupted. The outcome will be a pronounced daytime tiredness and frequent lapses into microsleep.

Risk factors contributing to sleep apnoea can be excess weight and alcohol.

Insomnia

Insomnia is understood to be a disturbance in falling and staying asleep, or a restless sleep. This can result in problems with concentration or an increase in daytime tiredness.

Risk factors are mentally-related disorders, such as depression.

Reasons Healthy People Sleep Poorly

- Mental strain or the inability to "switch off" due to personal problems, brooding, dystress
- Physical tension and restlessness
- Irregular sleeping hours related to flight duty
- Adverse sleeping environment (e.g. noise, heat, inconvenient layover hotel)
- Unfavourable sleeping habits (e.g. watching television in bed)
- Alcohol or medication

> The only effective treatment for fatigue is adequate sleep (Caldwell 1997).

The only efficient option available to compensate for a sleep deficit is to get several hours of continuous sleep. For not only is the quality of sleep, but also the quantity of sleep is of great importance when trying to compensate for a lack of sleep.

8.4 Influence of Alcohol and Medication

8.4.1 Alcohol

Although alcohol can have an altogether relaxing effect and, because of this, positively influence falling asleep, one should be careful to hold its consumption to a minimum and to not mistake it as a patent sleeping aid.

Alcohol suppresses REM sleep in such a way that mental restoration can take place only conditionally (Rosekind et al. 1995). Furthermore, stimulating substances are produced during the break down of alcohol that encourage premature awakening, thereby reducing the duration of sleep. As a general rule, the after-effects of alcohol felt the next day will also be performance inhibiting.

8.4.2 Medication

Even though the actual effect of sleep-inducing drugs as an aid to falling and staying asleep may be achieved, they oftentimes cause disruptions in the sleep structure. The effects can continue into the following day so that the person is neither fresh nor effective. Moreover, the risk of dependency—just as with the consumption of alcohol—should not be underestimated.

For this reason, the use of drugs to induce sleep is generally discouraged. If at all, then they should be taken only temporarily and under prescription from a physician trained in somnology. Permanent medication is difficult to reconcile with the fitness required for flying.

8.4.3 Melatonin

As already mentioned, melatonin is produced for sleep regulation by the body in the epiphysis (pineal gland) contingent on darkness. Therefore, it is natural to want to influence the internal clock by administering external melatonin.

Two effects have been attributed to this hormone: the encouragement of sleep and a shifting of the circadian phases. At first glance, these attributes seem to offer a solution to those persons working on shift schedules. However, the following must be considered at the same time:

An individual's personal circadian phases must be accurately known in order to realize a positive effect. Because melatonin is available in many countries as a food supplement, a proof of harmlessness has not yet been issued and precise dosage guidance has not been prescribed to date. This poses the risk that the body's own waking/sleeping rhythms will become completely confused and that the additional administration of melatonin will have serious counterproductive effects. This has been revealed by studies done on the effectiveness of taking melatonin (Sharkey et al. 2001). Deliberate light management, on the other hand, is considerably more effective than melatonin (see below).

> It is strongly recommended to avoid using melatonin for flight-related activities! (VC/DLR Alertness Management.

8.5 Fatigue-Related Factors

The following four factors have the greatest influence on fatigue and alertness:

- Time on task
- Time since awake
- Sleep deficit
- Circadian cycle

8.5.1 Time on Task

The longer one is involved with a task, the more tired he will become. Time on task defines the period of time during which a pilot carries out his duties (flight duty time or working time).

8.5.2 Time Since Awake

Time since awake (TSA) signifies the time that has passed from the moment a person awakens.

In a broad sense, the human organism can be compared to a highly complicated battery whose "state of charge" is influenced by a number of different factors. In addition to nutrition, oxygen and light, sleep represents the most outstanding restorative component. According to this model, the person "discharges" when he is not sleeping or resting. Beginning with the end of the sleep phase, a person's capacity for handling stress lowers as a function of the intensity and duration of the stress.

Insofar as a sleep deficit is not present, the point of time after waking at which fatigue will occur can be calculated using a simple mathematical model from the South Australian University (Dawson 2006). According to this model, TSA should be preferably less than the sum of the hours spent sleeping during the previous 48 h. Thereafter, the fatigue that develops will become too great. This establishes a good point of reference for determining whether one will be sufficiently fresh upon completion of the pending flight duty.

The calculation should be merely construed as a very rough indicator of the length of TSA, at the latest, at which significant fatigue should be anticipated. Under certain conditions, signs of fatigue may already emerge much earlier. The mathematical model does not distinguish between the individual influences that could interfere with the restoration process. These can be factors such as the segmentation of sleep or the state of sleep with relation to the internal clock. Furthermore, such influences as the time of day (body time) and stress density are not considered.

Example:

8 h of sleep during the first night + 4 h of sleep during the second night equal a total of 12 h. The point in time when fatigue can next be anticipated will be after these 12 h.

If a person wakes up at 7 a.m. in this example and adds up the time sleeping over the past 48 h, the resulting time will be that time at which fatigue can next be anticipated. In this case, 7 a.m. + 12 h = 7 p.m.

Thus, it will become critical for this cockpit crew if they are still on flight duty after 7 p.m.

Time since awake has a self-explanatory influence on our overall alertness. TSA is not regulated by law.

Causal factors related to fatigue and correlated to Time since awake are:

- Many pilots do not live in the immediate vicinity of the airport. In some cases, a commute of several hours prior to flight duty must be considered.
- Crew members in management functions often work even before reporting for flight duty.
- In particular, longer TSA's can be expected on late shifts and on night flights.

Rule of thumb for calculating fatigue	
	Wake-up time
+	Sleep duration over the previous 48 h
=	**Time of next fatigue onset**

Symptoms of fatigue		
Physical	**Mental**	**Emotional**
• Yawning • Heavy eyelids • Rubbing of eyes • Nodding of head • Microsleep	• Difficulty concentrating on the task • Reduced alertness • Difficulty remembering actions • Lack of communication about important matters • Lack of anticipation of future events • Inadvertent errors of comission • Inadvertent errors of omission	• More reserved or withdrawn than usual • Lack of energy • Lack of motivation to do well • Irritable behaviour towards collegues, family or friends

Fig. 8.2 Fatigue-related symptoms

8.5.3 Sleep Deficit

Sleep deficit occurs when a person gets less sleep then he actually requires (about 8 h per night). Reductions in performance can already be measured at 2 h below the normal individual length. Sleep deficit accumulates over the course of several evenings and can be diminished only by consequently trying to "catch up". In order to do this, it must be possible to complete at least one full 24 h waking/sleeping cycle, while two nights would be preferable (Rosekind et al. 1995).

8.5.4 Circadian Cycle

As previously mentioned, the circadian rhythm determines when a person is the most effective and when he is the least effective.

8.6 Fatigue-Related Symptoms

Fatigue expresses itself differently in each person. Many symptoms may appear suddenly in some people, while only a few symptoms may appear in others, yet to some extent more distinctly. Figure 8.2 provides an overview of fatigue's potential effects (Transport Canada 2007):

8.7 Fatigue and Performance

As mentioned earlier, a lack of sleep will significantly impact a person's performance capabilities, which will be greatly reduced due to this lack of sleep or sleep deprivation. This is already the case with a sleeping routine of five or fewer hours the night before flight duty (Dinges et al. 1997).

In the transport industry, in general, fatigue is the most common cause of accidents, exceeding even drug- and alcohol-related accidents (Akerstedt 2000). In the aviation industry, it is assumed that it plays a contributory role in about 4–7 % of all accidents (Kirsch 1996).

Cases have even been recorded where both pilots have fallen asleep in the cockpit at the same time (e.g. Guardian, 6 November 2007).

8.7.1 Workload and Flight Altitude

High cruising altitudes (= low air density) and a high workload are two further criteria that can intensify the slump in performance related to a lack of sleep (Hawkins 2002). This is particularly good to know for pilots flying in long-range operations, because high altitudes, time zone changes and a lack of sleep are factors they are frequently confronted with.

8.7.2 False Optimism

The so-called "false optimism" in connection with fatigue is very dangerous (VC/DLR Alertness Management). One's own performance capabilities may be inaccurately assessed by the individual in a similar manner as with dystress (Helmreich and Merritt 1998).

> Under the influence of fatigue, it is almost impossible for a person to correctly assess one's own performance capabilities in a subjective manner (VC/DLR Alertness Management).

This has been clearly demonstrated in the simulator with long-range crews being tested immediately following their flights, where risks were underestimated. Awareness of this potential misjudgement can be important in the case of a commander's decision to prolong the maximum duty time, for example, and a poll of the crew has resulted in their subjective assessment as to being sufficiently fit. Adequate risk management is possible only with knowledge of the origin and effects of fatigue. The rationale that "everything has gone fine up to this point" is inappropriate.

Table 8.1 Comparison between sleep deprivation and alcohol consumption

Sleep deficit (hrs.)	Comparative number of glasses of (U.S.) beer
8	10–11
6	7–8
4	5–6
2	2–3

8.7.3 Automation

Fatigue can also result from a "mental underload" such as when a lack of activity causes the eyes to "grow heavy". This so-called "boredom fatigue" will gain greater significance in the future with increased automation of the aircraft and procedures. According to one study (Colquhoun 1976), attention abates by up to 80 % within 1 h of undertaking monotone monitoring tasks.

8.7.4 Comparison of Sleep Deprivation and Alcohol

Table 8.1 shows a comparison between sleep deprivation and alcohol consumption with relation to performance (Roehrs et al. 2003). In this comparison, only response time is used as a benchmark, as this is slowed due to fatigue as well as due to alcohol consumption:

In other words: even after only one night of sleep deprivation (typical for long-range flights), response times are impaired to a greater degree than would be tolerated in road traffic due to alcohol consumption.

The National Transportation Safety Board (NTSB 2005) expresses it in a similar manner. A period of uninterrupted alertness of

- 16 h would result in a loss of performance capabilities corresponding to 0.5 per mille blood alcohol content.
- 22 h would result in a loss in performance capabilities corresponding to 0.8 per mille blood alcohol content

8.7.5 Chronic Sleep Deficiency

Sleep deficit is cumulative and can become chronic in nature. This can progressively lead to a permanent and linear impairment of performance that is barely noticeable to the individual (Gawron et al. 2001). Chronic sleep deficit and disruption can ultimately lead to an increased risk of illness in the gastrointestinal tract (Monk 1990).

8.7.6 Effects and Statistics

The consequences of fatigue in the cockpit can vary in type. An overview of the categories is provided below (VC/DLR Alertness Management, Dawson and Reid 1997):

Limited Cognitive Capabilities/Multi-Tasking

- Precision is degraded.
- Concentration and attention diminish; information processing becomes more difficult.
- Innovative thinking and flexible decision-making suffer.
- The capacity for integrating information is lost.
- A decision, once made, is revised less often.
- Lower standards of performance are accepted.

Target Fixation

- Tunnel vision
- Lack of flexibility
- Increased willingness to take risks

Motor Skills

- Reduced coordinating skills
- Worse timing
- Slowed reaction time

Communication

- The right words are more difficult to find
- The language used is less expressive

Microsleep

- Loss of orientation

Social

- Detached
- Some mistakes are accepted more readily
- One becomes less tolerant
- Ancillary tasks are neglected
- Persuasion becomes more difficult
- Confusion sets in more readily
- Uncomfortable

According to an NTSB study related to aircraft accidents experienced by large U.S. air carriers between 1978 and 1990, half of the captains (whose schedules were known) had been awake for over 12 h at the time of the accident. At the same time, half of the co-pilots had been awake for over 11 h. Overall, it could be observed that the crews who had been awake for an above average length of time made more work-related mistakes and more erroneous decisions than those crews whose Time since awake was below average (NTSB 1994).

Table 8.2 Captain duty hours and accidents by length of duty

Hours in duty period	Captain's hours	Exposure proportion	Accidents	Accident proportion	Accident proportion relative to exposure proportion
1–3	430,136	0.35	15	0.27	0.79
4–6	405,205	0.33	15	0.27	0.84
7–9	285,728	0.23	14	0.25	1.11
10–12	109,820	0.09	8	0.15	1.65
13 or more	12,072	0.01	3	0.05	5.62
Total	1,242,961	1.00	55	1.00	1.00
Calculated χ^2		14.89		10 % χ^2	7.8
Degrees of freedom		4		5 % χ^2	9.5

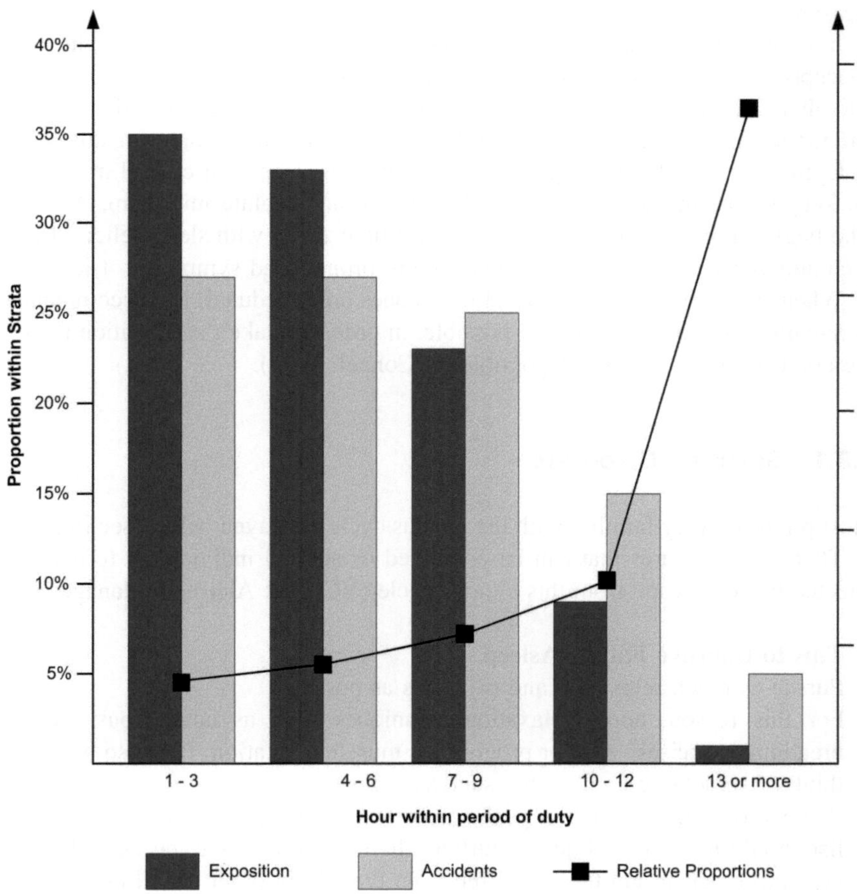

Fig. 8.3 Captain duty hours and accidents by length of duty

Lengthy pilot duty times do not only indicate a greater potential for error, but they consistently point to a greater potential for accidents, as well.

The results of another study reveal that there is an obvious connection between the probability of accident occurrence and the length of pilot flight duty times (see Table 8.2). Despite the fact that only 10 % of all data records analysed in this study had flight duty times greater than 10 h, 20 % of the human factors-related accidents occurred during these flights (Goode 2003).

The graph below is derived from this table (see Fig. 8.3) and depicts the probability of an accident occurrence with increasing flight duty times:

8.8 Fatigue and Alertness Management

The effects of fatigue reveal themselves in many different forms, so there is no "basic recipe" for to handle them. There are measures that can be taken to successfully cope with fatigue, however, which can be helpful by themselves or in combination with other measures. These are distinguished between preventive measures and aids for flight activities.

It is absolutely essential that airlines integrate training related to the basic concepts of fatigue and sleep, or performance, as well as alertness management into their CRM seminars. The Vereinigung Cockpit has developed a one-day Fatigue seminar together with the DLR that serves as an example for this.

Optimization of both the quality and quantity of sleep is of critical importance for keeping the impact of fatigue and jet lag to an absolute minimum. A person who begins a trip scheduled to cross several time zones with sleep deficit already accumulating will generally experience more pronounced symptoms of jet lag.

When short trips crossing several time zones are scheduled, it is recommended to remain on the original time, if possible, in order to make the transition back to the point of origin as easy as possible (O'Connell 1998).

8.8.1 Sleeping Disorders

Most pilots are very familiar with the vicious cycle portrayed below (see Fig. 8.4).

There are measures that can be combined or applied individually to help any effected person escape from this vicious cycle (VC/DLR Alertness Management).

Tips to Improve Falling Asleep

- Pursue as much relaxation and calmness as possible.
- For this reason, apply relaxation techniques such as autogenous training, breathing exercises, yoga or progressive muscle relaxation. It is also helpful to think of something comforting, such as a pleasant vacation memory.
- Write a to-do list for the things that need to be accomplished the next day. The list should help discard any disturbing thoughts at bedtime, because the head needs to be clear and the mind free from thought in order to fall asleep.

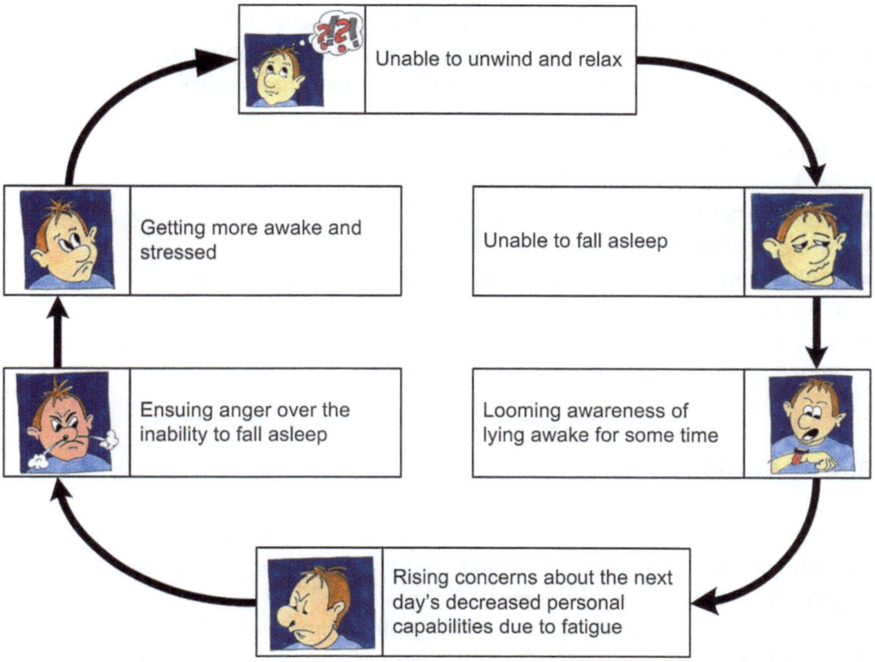

Fig. 8.4 Vicious cycle of sleeping disorders (© DLR)

Interestingly: studies have shown that people with sleeping disorders underestimate their actual sleep duration.

8.8.2 General Measures for a Sound Sleep

Measures facilitating a sound sleep can be divided into two categories: preventive strategies and operational strategies (NASA 2002). Preventive strategies can further be distinguished between general and flight duty-specific measures, depending upon whether the individual is anticipating work duty.

Preventive Measures: General

- Healthy nutrition
- Sufficient exercise
- Adequate sleeping environment (sleep hygiene), such as a darkened room, a suitable room temperature, etc.
- Maintenance of a regular waking/sleeping rhythm.
- Develop and maintain a regular routine prior to going to bed.
- Eat light meals when hungry and avoid heavy foods.
- Avoid alcohol and caffeine prior to going to bed.
- Employ physical and mental relaxation exercises.
- If it is not possible to fall asleep within 30 min, then get out of bed again.

Preventive Measures: Prior to and During Flight Duty

- Getting a potential 8 h of sleep prior to a workday is optimal.
- Take targeted naps. A nap can temporarily raise attentiveness.

Operational Measures

- Seek out conversation with others.
- Physical activity: stretching, exercising, chewing gum
- Strategic caffeine consumption
- Light management in the cockpit can influence crew alertness.
- Give heed to nutrition
- When it can no longer be avoided or is no longer doable: interrupt the flight duty

8.8.3 Caffeine

Caffeine is an agent frequently used to combat fatigue during shift duty. Yet, caffeine cannot compensate for a lack of sleep but merely provides a temporary suppression of the signs of fatigue.

As a rule, the intake of at least 200 mg of caffeine every 2 h is recommended (Caldwell and Caldwell 2003). The amount will vary, however, depending on the person's individual tolerance to caffeine.

Some tips with regard to the consumption of caffeinated beverages:

- Coffee usually begins to work after 30 min. Half of the caffeine will be depleted after 3–5 h, but this, too, will vary with the individual.
- Coffee should no longer be consumed within about 3 h prior to going to bed in order to avoid interference with falling asleep or disrupting the quality of sleep.

Table 8.3 provides an overview of the amount of caffeine contained in popular beverages (Transport Canada 2007):

Figure 8.5 shows how the volume of caffeine in the blood varies over time when ingested with coffee (VC/DLR Alertness Management). Generally speaking, it can be said that:

- Coffee takes effect after approx. 30 min.
- Half of the caffeine will be depleted after approx. 3–5 h.

8.8.4 Napping

Napping is an effective way to reduce fatigue. With consideration given to the text below, the DLR emphatically recommends engaging in planned and coordinated naps in the event of a heightened onset of fatigue during cruise flight in order to be at the highest possible level of proficiency for the approach and landing. Naturally, this applies only if napping during flight operation has been regulated by means of an official Napping policy.

Here is an excerpt from the legislative documentation:

> EU-OPS 1.310 Crew members at stations(a) Flight crew members(3) During all phases of flight each flight crew member required to be on flight deck duty shall remain alert. If a lack of alertness is encountered, appropriate counter-measures shall be used. If unexpected fatigue is experienced, a controlled rest procedure, organised by the commander, can be used if workload permits.

Brooks and Lack published the following graph in their study (see Fig. 8.6), which illustrates the correlation between napping and performance:

Napping not only serves as a preventive measure, but as an operational measure, as well. On the one hand, naps can be used as a prophylaxis in order to ensure that performance capabilities are available at a later point in time. On the other hand, napping can also reduce the Time since awake and the Time on task, thereby preserving performance capabilities over a longer period of time (Rosekind et al. 1995). This does not mean that napping generates a "sleep credit" or that "pre-sleeping" is a valid method for warranting safe flight operations, however, (Dinges et al. 1988).

There are two negative aspects associated with napping: sleep inertia and the potential negative impact napping may have on subsequent periods of sleep.

It is possible to fall into a deep sleep phase with a nap whose sleep phase is longer than 20 min in duration. After waking, there is also the danger associated with sleep inertia; a state in which more than 15 min may be needed to orient one's self and to be able to work with full awareness once again. This can be problematic in situations such as emergency ensuing during an in-flight nap.

Table 8.3 Caffeine content in popular beverages

Coffee (250 ml)	Tea (250 ml)	Caffeinated beverages
Instant 65–100 mg	Green tea 8–30 mg	Cola 50 mg
Fresh brewed 80–135 mg	Normal tea 50–70 mg	Red bull 80 mg
Espresso 100 mg		

Fig. 8.5 Volume of blood caffeine over time when ingested with coffee

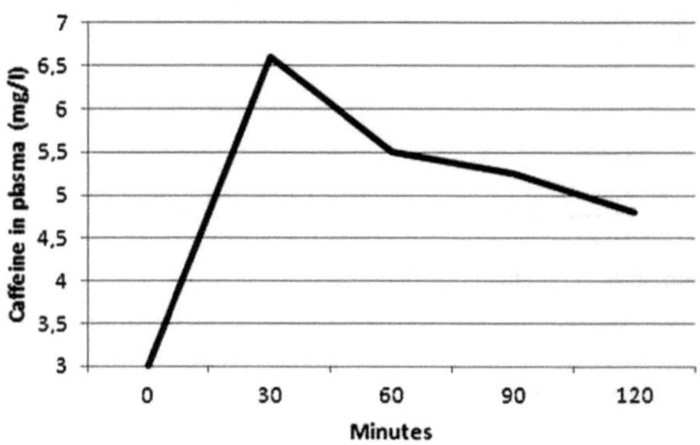

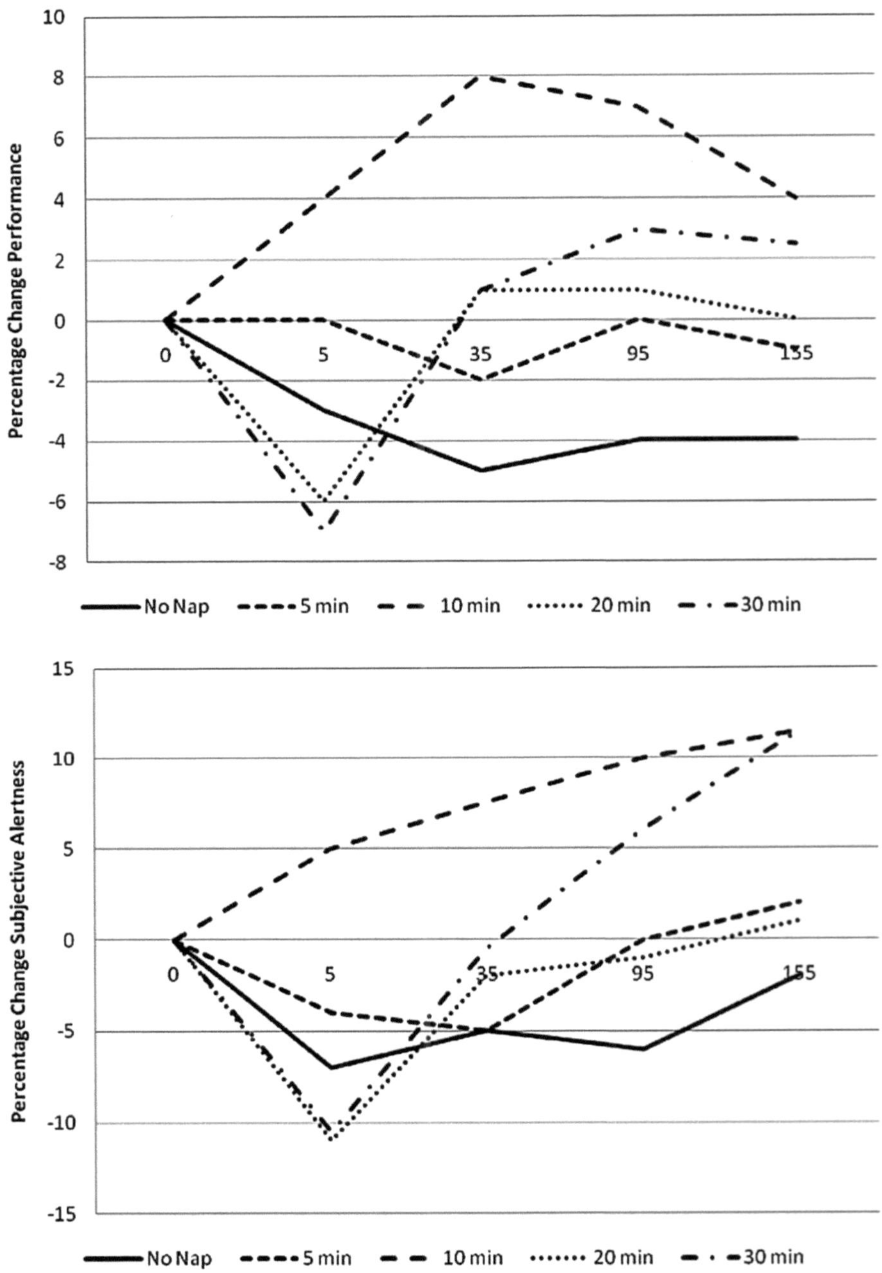

Fig. 8.6 The effects of napping

A long nap at an unfavourable time can result in not being able to fall asleep at, or to sleep through the usual time.

Recommendations for Effective Napping

- Napping should be dealt with restrictively and called upon only as a last resort. Naps should not only be coordinated between the cockpit personnel, but with the cabin crew, as well, provided the opportunity. Flight operations should describe the necessary communication required with the cabin crew in its napping policy in order to prevent all cockpit crew members from falling asleep at the same time—unnoticed. Otherwise, napping should be strictly prohibited.
- The actual sleep phase of a nap should not exceed 10 min. This means that, for all intents and purposes, the overall length of the pause for napping should last 30–45 min. Several minutes will be needed for falling asleep and about another 15 min should be planned following the nap to ensure that the person has passed out of the sleep inertia phase and is fully functional once again (Dinges et al. 1991).
- Whenever possible, naps should be taken early, meaning before one senses the clear signs of fatigue.

8.8.5 Light Management

Light and darkness have a direct influence on the circadian rhythm by way of the melanopsin pigment in the ganglion cells located in the retina of the eye. This can be influenced to some degree through deliberate light management, thereby governing whether one remains in the rhythm of the departure airport or, in contrast, more quickly adapts to the rhythm of the destination airport.

On westbound flights with short layovers, for instance, it is common to want to remain in the rhythm of the departure airport. In such cases, it would be sensible to retire to the hotel room at the normal sleeping time in the east, while it is still light out in the west, and to darken the room to make it as light-proof as possible.

In a similar manner, it would be helpful on eastbound flights to sojourn in bright rooms long past the onset of local darkness. Then, at the normal sleeping time, during the local early morning hours, the surroundings should be darkened when retiring to bed. It is important, therefore, that the hotel room be darkened and as light-proof as possible to prevent being prematurely awakened due to scattered light.

If a person deliberately wants to adapt quickly to the new day/night rhythm, such as might be the case with an eastbound flight, then it would be helpful to wake up early and to spend some time in the daylight.

This deliberate form of light management has proven itself to be far superior to the artificial supply of melatonin (Dawson et al. 1995).

It is not yet clear, however, just how much light is necessary for regulating the circadian rhythm. Studies have shown that bright room lighting (1,000–2,000 Lux) already has an effect (Phipps-Nelson et al. 2003). Stronger light, such as would be found outdoors on a cloudy day (10,000 Lux) is preferable. The decisive factor is the wavelength of the light. It should be in the blue light range with a length of 425 nm. A transportable blue light lamp, which emits precisely the wavelength required for sleep management, is available from Philips (Philips goLite BLU).

If a sleep deficit exists or if the individual is at a circadian low point, then light will be less effective as a wakeup aid. It may produce a temporary alarming effect, but it cannot compensate for the underlying problem of fatigue and its associated effects.

For long-range crews, this could mean: The presence of mixed signals occurring simultaneously because of time zone crossings, such as (1) internal clock = night, (2) surroundings = daylight, may result in an internal desynchronization. The consequence is fatigue.

Fatigue can occur on a morning flight such as when returning from Asia to Europe. Little sleep was possible during the previous Asian night due to the body's rhythm. At the same time, the takeoff and subsequent cruise flight take place in bright sunlight, despite the fact that the body is now programmed for sleep.

The use of bright lighting in the cockpit can be helpful during night flights. But beware: under these conditions it is practically impossible to distinguish other traffic. This is particularly unacceptable in an environment where it is anticipated that the other aircraft will have unreliable TCAS equipment. Lightening associated with thunderstorms will also be barely distinguishable, making it easier to inadvertently penetrate large cumulus clouds where severe turbulence can be expected.

8.8.6 Alertness Planning

As described above, fatigue will depend on many factors: The amount of sleep on the day prior to duty, Time since awake, Circadian rhythm and Time on task.

Computer programs are now available that are capable of recording these factors, and which enable reliable forecasting of the fatigue that can be expected at the end of a person's pending flight duty. The programs allow the following to be determined:

- Will a person still be sufficiently alert towards the end of the pending flight duty he is about to commence upon? Such information provides the individual with a solid basis for making a decision about removing himself from flight duty, even before reporting for that duty.
- Pause-planning on long-range flights with an increased crew contingent is simplified. The captain, as the person responsible for the flight, should not draw back from taking the pause that "best" fits his needs in order to be as fit as possible for the approach and landing. Accordingly, the pilot scheduled to sit at the rear during landing should take the least favourable pause (VC/DLR Alertness Management).

- Potential decisions as to nap-planning are made easier because it is known ahead of time whether they will actually be beneficial for the landing.
- Decision-making support is available for planning the workload during approach. Depending on the state of fatigue, for example, visual or other high stress-related approaches should be refused by the operating flight crew member subjected to it.

One program that has been introduced and is being used by the U.S. Air Force and the U.S. Federal Railroad Administration is the Fatigue Avoidance Scheduling Tool (FAST^{TM}) (Mallis et al. 2004). This program uses an accelerometer or so-called Actigraph worn on the body to reliably measure when and how much a person sleeps.

The U.K. Health and Safety Executive (HSE) has developed a similar program, in which fatigue is set in relation to the probability of accident occurrence (Spencer et al. 2006). The Fatigue/Risk Calculator can be downloaded at the HSE website: www.hse.gov.uk (status: May 2009).

The DLR has tested the ALERT program, which is tailored to the aviation industry and makes possible more precise statements specifically for pilots. The Vereinigung Cockpit participated in the further development of this program as well as its adaptation into a flight log. We recommend its usage upon release within short time (status: June 2012) most probably in collaboration with Lufthansa Systems.

8.9 Summary

- The human being is the only component in the aviation industry that does not function reliably around the clock.
- Fatigue is a contributing factor in 4–7 % of all aviation-related accidents, particularly those related to long-range and night flights.
- The probability of an accident occurrence increases considerably with flight duty times longer than 10 h and even more significantly with flight duty times longer than 13 h.
- Flight crews require knowledge about sleep, fatigue and performance.
- Flight crews must do everything possible in order to remain resilient, especially at the end of their flight duty day.
- Flight crew members over 50 years of age have a greater problem with fatigue and sleep deficit than younger flight crew members.
- Flight crews should deliberately pursue sleep hygiene and know how to escape the vicious cycle of sleeping disorders.
- The reasons for poor sleep should be well known.
- The use of alcohol, medication and melatonin to improve sleep should be discouraged.

- Flight crew members must be aware of the influence of the TSA on individual performance capabilities (commuting, commitments prior to reporting for flight duty, etc.).
- Low air densities, high workloads and/or underloads are tiring.
- The degree to which a person is fatigued is almost always underestimated. This makes it easier for a person to be lured into taking avoidable risks.
- The reduction of performance even after only one night with less than 5 h of sleep is considerable.
- When signs of fatigue begin to appear or when it becomes unavoidable while in the cockpit:
- Seek out discussions and conversations
- Get some exercise (even chewing gum can help)
- Deliberately consume caffeine
- Consider napping
- With reliable TCAS equipment and a safe distance from thunderstorm cells: Switch on the cockpit lighting at night
- Ultima ratio: discontinue flight duty as a means of last resort, if possible
- Acquire and consciously practice light management during layovers and at home.
- Use an appropriate computer program for alertness management where available.

References

Akerstedt T (2000) Consensus statement: fatigue and accidents in transportation operations. J Sleep Res 9:395

Caldwell JA (1997) Fatigue in the aviation environment: an overview of the causes and effects as well as recommended countermeasures. Aviat Space Environ Med 68:932–938

Caldwell JA, Caldwell JL (2003) Fatigue in aviation. Ashgate, Burlington

Colquhoun P (1976) Psychological and psychophysiological aspects of work and fatigue

Dawson D, Encel N, Lushington K (1995) Improving adaptation to simulated night shifts: Timed exposure to bright light versus daytime melatonin administration. Sleep 18:11–21

Dawson D, Reid K (1997) Fatigue, alcohol and performance impairment. Nature 388:23

Dawson D (2006) Defences in Dept., Lecture at the flight safety foundation EASS, Paris

Dinges DF, Pack F, Williams K, Gillen KA, Powell JW, Ott GE, Aptowicz C, Pack AI (1997) Cumulative sleepiness, mood disturbance, and psychomotor vigilance performance decrements during a week of sleep restricted to 4–5 hours per night. Sleep 20:267–277

Dinges DF et al (1991) Preplanned cockpit rest: effects on vigilance performance in long haul flight crews. Aviat Space Environ Med 62(5):451

Dinges DF, Whitehouse WG, Orne EC, Orne MT (1988) The benefits of a nap during prolonged work and wakefulness. Work Stress 2:139–153

EASA, Definition Fatigue: GM OR.OPS.025.FTL

Gander PH, Myhre G, Graeber RC, Andersen HT, Lauber JK (1989) Adjustment of sleep and the circadian temperature rhythm after flights across nine time zones. Aviat Space Environ Med 60(8):733–743

Gawron VJ et al (2001) An overview of fatigue. In: Hancock PA, Desmond PA (eds) Stress, workload and fatigue. Lawrence Erlbaum Associates, Mahwah, pp 581–595

Goode JH (2003) Are pilots at risk of accidents due to fatigue?. Federal aviation administration office of aviation policy and plans, Washington
Hawkins F (2002) Human factors in flight. Ashgate publishing, Farnham
Helmreich FH, Merritt AC (1998) Culture at work in aviation and medicine. Ashgate, Aldershot
Kirsch AD (1996) Report on the statistical methods employed by the U.S. FAA and its cost benefit analysis of the proposed "flight crew member duty period limitations, flight time limitations and rest requirements". Comments of the Air Transport Association of America to the FAA notice 95-18, FAA docket number 28081, Appendix D, pp 1–36
Mallis MM, Mejdal S, Nguyen TT, Dinges DF (2004) Summary of the key features of seven biomathematical methods of human fatigue and performance. Aviat Space Environ Med 75:A4–A14
Monk TH (1990) The relationship of chronobiology to sleep schedule and performance demands. Work stress 4:227–236
NASA (2002) Crew factors in flight operations XV: alertness management in general
NASA/TM-2002-211394, Rosekind et al
NTSB (1994) Safety study: a review of flight crew involved. Major Accidents of U.S. Air Carriers, 1978-1990. PB 94-917001 NTSB/SS-9401, Washington
NTSB (2005) Course, investigating human fatigue factors, NTSB academy
O'Connell D (1998) Jetlag, how to beat it. Ascendent, London
Phipps-Nelson J, Redman JR, Dijk DJ, Rajaratnam SM (2003) Daytime exposure to bright light, as compared to dim light, decrease sleepiness and improves psychomotor vigilance performance. Sleep 26:695–700
Roehrs T et al (2003) Ethanol and sleep loss: a dose comparison of impairing effects. Sleep 26:981–985
Rosekind MR, Smith RM, Miller D, Co EL, Gregory KB, Webbon LL, Gander PH, Lebacqz JV (1995) Alertness management: strategic naps in operational settings. European Sleep Research Society
Samel A, Wegmann HM, Vejvoda M (1997) Aircrew fatigue in long-haul operations. Accident Anal Prev 29(4):439–452 (Elsevier Science Ltd)
Sharkey KM, Fogg LF, Eastman CI (2001) Effects of melatonin administration on daytime sleep after simulated night shift work. J Sleep Res 10(395):181–192
Spencer MB, Robertson KA, Folkert S (2006) The development of a fatigue/risk index for shift-workers. HSE Books, London
Strauss S (2009) Pilot fatigue.http://aeromedical.org/Articles/Pilot_Fatigue.html. Accessed on 25 Aug 2009
Transport Canada (2007) An introduction to managing fatigue. TP 14572E

Recommendations by the German Airline Pilot's Association

9

CRM holds a position of high importance in the avoidance of risk and the corresponding safe flight conduct. For this reason, it should be a part of a pilot's core competencies right alongside his piloting, technical and procedural skills.

CRM is comprised of general fundamental information with operator-specific content and design. The requirements for CRM training are based on EU-OPS 1.965, with details specified in TGL 44.[1] Furthermore, EU-OPS 1.965 Annex 1 (a) (4) contains a list of the CRM sections that must be covered during recurrent training (see Annex 1 of this chapter).

The EASA plans to cede the implementation of these guidelines to the airlines on a voluntary basis. The German ALPA (Vereinigung Cockpit "VC") rejects this position. The structure of the training sections as defined in AMC EU-OPS 1.965 should be retained in the training program for safety reasons.[2]

In this book, the VC has provided an exemplary description of the type-independent section content requirements. We recommend structuring CRM training in three different steps:

- The required general learning objectives should be configured in written format, such as within the framework of a flight operations handbook for line operations and as a manual for flight schools.
- CRM seminars have established themselves as a bridge between the theory and its practical application in the daily work routine. This is so because the individual can practice the fundamental parts of CRM built upon the general

Vereinigung Cockpit e. V, Main Airport Center,
Unterschweinstiege 10, 60549 Frankfurt, Germany
e-mail: office@vcockpit.de

[1] TGL44, p. 141 ff. version from 1 June 2008.

[2] AMC OPS 1.943/1.945(a)(9)/1.955(b)(6)/1.965(e) Crew Resource Management (CRM), Para. 6 Implementation of CRM, see Annex 2.

H.-J. Ebermann and J. Scheiderer (eds.), *Human Factors on the Flight Deck*,
DOI: 10.1007/978-3-642-31733-0_9, © Springer-Verlag Berlin Heidelberg 2013

learning objectives without being distracted by other influencing factors associated with that daily work routine. CRM seminars should be built upon one another (see Annex 2 of this chapter). Accordingly, the German Airline Pilot's Association has developed a model seminar addressing the subjects of leadership and teamwork while demonstrating how we envision structuring the material therein as part of this bridging function.

- CRM content in line with this book should be taught in and applied to all instructional, training and check events in the simulator and the aircraft. An assessment should be made with the exception of simulator checks (FCL, OPC). CRM feedback related to any training measures should always be submitted. Video feedback with an appropriate debriefing is beneficial for simulator sessions.

Accordingly, every chapter in this book is comprised of a general section, whose subject matter should become part of the theoretical learning objectives and CRM seminars. Specific CRM behaviour patterns can be subsequently found in most chapters, where reasonable, which can serve as a basis for the simulator and aircraft syllabi. In addition, assessment sheets can also be integrated. It is incumbent upon the airlines to undertake the detailed implementation of the statutory requirements related to CRM training and CRM assessment.

Training is more likely to achieve its desired goals when its effect is assessed. Albeit, CRM assessment also conceals risks:

- Abuse on the part of the examiner or his principal client.
- Inadequate organization of the general framework can have a negative influence on the training results. Furthermore, the IFALPA fears the fabrication of a "put-on" behaviour without a genuine change in behaviour.[3]

In order to avoid these risks, the VC believes the following to be essential for optimal CRM training and CRM assessment:

- The subject areas to be assessed must be described, published and trained.
- The additional training measures for examiners should be comprehensively set forth and carried out.[4] The VC considers the "framework program for the observation and assessment of CRM skills", as set forth by the LBA (German civil aviation authority), to be a minimum requirement.
- Only captains with the status of TRE/TRI/SFE are to be appointed as examiners, whereby the check is to be carried out from the observer's seat. The

[3] IFALPA Annex PILOT PROFICIENCY CHECKS, 9.4.4.1 CRM: When first introduced, a cornerstone in the acceptance for CRM training was the assurance that CRM should be without checking. Much of the value and strength of CRM is based on this principle. The introduction of a checking or assessment process has the potential to destroy such benefits. IFALPA therefore opposes an assessment process of any nontechnical aspect.

[4] CAA Guide to Performance Standards for Instructors of CRM training in Commercial aviation, Riverprint Ltd, Sept. 1998.

assessment system must satisfy the validity,[5] reliability[6] and, in particular, objectivity[7] criteria, so that the various examiners arrive at the same conclusions.
- A Failure policy should be established,[8] whereby a deficit in CRM shall result in a training measure designed to re-establish CRM competency.

Based on the JAR-TEL NOTECHS System,[9] the following principles are to be maintained during any assessment:

Need for Technical Consequences
Deficient/incorrect CRM behaviour, alone, should not lead to a "failed check", but should merely be used to explain limit transgressions or other flight hazards that, in themselves, would result in a failed check.

Explanation Required
Particularly in the event of a negative CRM behaviour assessment, the examiner is obligated to record the elements and categories of the deficiency in writing, as well as to provide a description of the consequences for flight conduct.

Repetition Required
If CRM is to result in a negative assessment, then it shall not be based on the observation of an individual incident, but rather on a corresponding observable behaviour pattern.

Only Observable Behaviour
Only observable behaviour should be appraised, but not the personality or the emotional attitude of the pilot.

Rating Scale
A breakdown of the assessment that extends above and beyond the use of the two point rating scale criteria of "acceptable"or "unacceptable"is necessary. This is in order to be able to carry out and improve analyses and trend observations of anonymous data from the training system.

9.1 Annex 1

EU-OPS English, Appendix 1 to JAR–OPS 1.965(a) (4) Recurrent training and checking—Pilot's Crew Resource Management (CRM)

1. Elements of CRM shall be integrated into all appropriate phases of recurrent training; and

[5] What needs to be measured will be measured.

[6] The same result is obtained when repeated.

[7] Another examiner arrives at the same conclusions.

[8] AMC OPS 1.943/1.945(a)(9)/1.955(b)(6)/1.965(e) Crew Resource Management (CRM), Para 8.3: Operators should establish procedures to be applied in the event that personnel do not achieve or maintain the required standards.

[9] NLR-TP-98518, Hoofddorp 1998.

2. A specific modular CRM training programme shall be established such that all major topics of CRM training are covered over a period not exceeding 3 years, as follows:
   (A) Human error and reliability, error chain, error prevention and detection;
   (B) Company safety culture, SOPs, organisational factors;
   (C) Stress, stress management, fatigue and vigilance;
   (D) Information acquisition and processing, situation awareness, workload management;
   (E) Decision making;
   (F) Communication and coordination inside and outside the cockpit;
   (G) Leadership and team behaviour, synergy;
   (H) Automation and philosophy of the use of Automation (if relevant to the type);
   (I) Specific type-related differences;
   (J) Case based studies;
   (K) Additional areas which warrant extra attention, as identified by the accident prevention and flight safety programme (see JAR-OPS 1.037).

9.2 Annex 2

AMC OPS 1.943/1.945(a)(9)/1.955(b)(6)/1.965(e) Crew Resource Management (CRM), Para 6 Implementation of CRM

Core elements	Initial CRM training	Operator's conversion course when changing type	Operator's conversion course when changing operator	Command course	Recurrent training
(a)	(b)	(c)	(d)	(e)	(f)
Human error and reliability, error chain, error prevention and detection	In depth	In depth	Overview	Overview	Overview
Company safety culture, SOPs, organisational factors		Not required	In depth	In depth	
Stress, stress management, fatigue and vigilance			Not required		
Information acquisition and processing situation awareness, workload management		Overview			
Decision making					
Communication and co-ordination inside and outside the cockpit			Overview		
Leadership and team behaviour synergy					
Automation, philosophy of the use of automation (if relevant to the type)	As required	In depth	In depth	As required	As required
Specific type-related differences			Not required		
Case based studies in depth	In depth	In depth	In depth	In depth	As appropriate

Glossary

1000 ft marker The ideal touchdown point for a landing aircraft, 1000 feet ($\sim$300 m) beyond the runway threshold

Abnormal (procedure) A standard procedure to follow in the event of the failure of an aircraft component, such as an engine

Airway An air route published by air traffic control

Alerts A low priority, visual or aural warning triggered by the aircraft's on-board monitoring system

Alternate A diversion airport when the approach to the destination airport cannot be completed

Altitude alert An aural signal when approaching a planned flight altitude or when about to overshoot it

Approach An approach procedure to an airport

Assertiveness The self-assertion or the capacity for enforcement that is absolutely necessary for critical feedback

ATC controller A flight controller (ATC = Air Traffic Control)

Aural warning An acoustic warning signal

Auto speedbrake Spoilers that extend automatically upon landing to provide air braking action

Autothrottle Automatic engine thrust regulation

Base check A semi-annual check flight prescribed by the authorities for maintenance of the airline transport pilot's license

Basic flying The manual flying of a transport aircraft without the support of automatic systems

Boarding The process through which passengers enter the aircraft

Cabin briefing The briefing given to the cabin crew by the captain at the start of their common flight duty

H.-J. Ebermann and J. Scheiderer (eds.), *Human Factors on the Flight Deck*,
DOI: 10.1007/978-3-642-31733-0, © Springer-Verlag Berlin Heidelberg 2013

Callout Precisely defined calls by a pilot for such events as a deviation from a planned flight profile or a change in the aircraft's automatic control system mode.

Call sign The call signal assigned to a ground station or an aircraft to be used during radio communication

CAT III approach Precision instrument approach according to precision category III

Catering In-flight refreshments and meals

CAVOK Clouds and visibility okay; an expression used in aviation to identify unproblematic visibility and cloud ceilings

Centerline The middle line of the runway

CFIT Controlled Flight into Terrain; an accident category, in which a fully functional aircraft makes unintentional contact with the terrain, water or an obstacle

COMM Abbreviation for Communication; the technical communication between aircraft/ground station

Commitment to stay The point in time at which a diversion to the alternate airport is no longer possible due to a lack of fuel

Convective weather Weather comprised of cumulus or cumulonimbus clouds; represents a source of severe turbulence and wind shear

Copilot The captain's deputy; the captain bears overall responsibility for the flight while a copilot bears responsibility merely for his own actions.

CPT Captain

Crew Resource Management (CRM) The effective use of resources by the crew in order to ensure the safe and efficient operation of the aircraft.

Debriefing A briefing that takes place at the end of a flight or flight duty

Delay vector An alternative routing assigned by air traffic control, such as when is needed to establish an arrival sequence

DME Distance Measuring Equipment; an on-board instrument for measuring distance to a radio navigation aid

EFIS Electronic Flight Instrument System; depiction of the most important flight guidance instruments on a display monitor

FDM Flight Data Monitoring; the pro-active use of flight data for accident prevention

Flag A warning marker signalling the loss of a cockpit indication

Flap setting The flap positions used for takeoff and landing

Flight Envelope Protection A protection system that prevents the aircraft flight operating limits from being exceeded (e.g. speed, flight attitude)

Fly-by-wire The transmission of control signals to the control surfaces by electronic, and not by mechanical means

FMA Flight Mode Annunciator; a cockpit display of the individual autopilot and Autothrottle control modes

FMS Flight Management System; an on-board navigation, flight performance and flight control computer

FO First officer; synonymous with copilot

FORDEC The decision model for complex decisions; details can be found in the chapter on Decision making

Fuel dumping The discharging of fuel in order to reduce the aircraft weight for landing

Gear Interlock System A system that prevents an unintentional movement of the landing gear

Go-around Synonymous with missed approach; the missed approach procedure when a landing is not possible

Ground Proximity Warning System A system providing timely reports when approaching ground proximity or an obstacle

Heading The horizontal direction of flight

Holding Flight procedure for in-flight loitering

Human factor A general term for anything that has to do with the improvement of human performance at the workplace. With respect to aviation, it is concerned with the ergonomics, process optimisation, crew resource management and related training measures deemed necessary for the prevention of human error and the optimization of flight safety. This comprises all participants within the aviation industry: crews, aircraft manufacturers, maintenance facilities, the airlines and air traffic control.

ICAO International Civil Aviation Organisation; an agency of the United Nations (UN) that works to ensure commonality of worldwide aviation standards. The ICAO is comprised of 180 member states.

IFR Instrument Flight Rules; the rules governing flight in instrument meteorological conditions

ILS Instrument Landing System; the most common and most precise radio-supported approach aid with a horizontal and a vertical localiser beam

Limitation The limiting values for technical systems or flight parameters, such as approach speeds, maximum descent rates when close to the ground, etc.

Line check A flight check of airline pilots performed annually in normal flight operating conditions in accordance with EU-OPS.

Line-up The instruction to taxi into position on the runway

LOFT Line Oriented Flight Training; simulator training that replicates an entire airline flight or airline flight situation as realistically as possible in real time.

Maintenance The techniques and procedures used to maintain an aircraft

MCP Mode Control Panel; equipment for entry inputs into the Auto Flight System

MEL Minimum Equipment List; a document published by the aircraft manufacturer listing operating constraints and recommendations for flight operations with system failures that are not immediately repairable.

Minimum The lowest visibility and cloud ceiling established for instrument approaches

Missed approach Synonymous with go-around; the missed approach procedure when a landing is not possible

Movable throttle The cockpit thrust lever that moves automatically during Autothrottle operations

Near miss The near-collision of two aircraft while in the air

Need-to-know items Knowledge that may not be easily looked up in the handbooks under adverse circumstances; it must be committed to memory

NOTAM Notice to Airmen; a standardized ICAO format for disseminating safety-relevant information to airmen

Off blocks The point in time at which an aircraft moves under its own or external power for the commencement of taxiing for flight

PAPI Precision Approach Path Indicator; an optical display of the correct approach angle

Pilot flying The pilot flying; the crew member actually controlling the aircraft

Pilot not flying/Pilot monitoring The pilot not flying; the crew member not actually controlling the aircraft, but who handles radio communications and the required supplemental tasks while monitoring the Pilot Flying, addressing and providing corrective assistance for his work-related errors.

Poor airmanship A prevalent, but not clearly defined term related to the insufficient capacity for piloting an aircraft

R/T Radio Telephony; radio communication

Ramp agent The responsible person for operationally-related ground activities

Readback The reading back of a message transmitted through radio communication in order to eliminate misunderstanding

Refresher Regular training events in the simulator to practice system failures and emergency situations

Roster The rhythm and (minimum) separation of flight duty and rest periods within the pilots' duty plans

Situational awareness Awareness of one's overall situation; a term used for the timely acquisition of all parameters required for the conduct of a flight

Slot A takeoff time issued by Eurocontrol or by an individual airport in order to avoid waiting periods on the ground and in the air

Speed clacker An aural warning triggered when the maximum speed of an aircraft is violated

Squawk A numerical digit sequence transmitted continuously to other aircraft and to air traffic control by an on-board radio transponder for identification purposes

Standard Operating Procedures (SOP's) Standardized procedures for all types of operational processes in order to facilitate safe and efficient operations. SOP's are helpful in avoiding non-essential arrangements between the pilots related to operational processes, among other things. They are established by the aircraft manufacturers and operators, and are not subject to voluntary compliance but must be implemented in a precise manner.

Start-up request A request made to air traffic control for the issuance of permission to start engines

Sterile Cockpit A policy defined in the flight operations handbook prohibiting private conversations and activities during critical phases of flight in order to avoid distraction

Strobe lights Flashing lights on the tail and wingtips that serve to enhance a more rapid visual recognition of the aircraft

TCAS The name given by a component manufacturer for its ACAS (Airborne Collision Avoidance System); a system for the prevention of mid-air collisions between aircraft

Term Meaning

Three-pointer altimeter The conventional mechanical or electronic method of barometric altitude indication with three needle pointers, and which can be easily misinterpreted

Touch-and-go A landing immediately followed by a takeoff, commonly used for flight and landing training

VASIS Visual Approach Slope Indicator System; an optical display of the correct approach angle

VFR Visual Flight Rules; the rules governing flight in visual meteorological conditions

VOR VHF Omnidirectional Radio Range; a navigational radio beacon with which bearings can be established by an on-board receiver in the aircraft

Warnings Optical, aural or tactile warnings triggered by monitoring systems on-board the aircraft and that are of a higher priority than alerts

Index

H.-J. Ebermann and J. Scheiderer (eds.), *Human Factors on the Flight Deck*,
DOI: 10.1007/978-3-642-31733-0, © Springer-Verlag Berlin Heidelberg 2013

Hans-Joachim Ebermann
Joachim Scheiderer
Editors

Human Factors on the Flight Deck

Safe Piloting Behaviour in Practice

Springer

Editors
Hans-Joachim Ebermann
Main Airport Center
Vereinigung Cockpit e.V.
Frankfurt
Germany

Joachim Scheiderer
Main Airport Center
Vereinigung Cockpit e.V.
Frankfurt
Germany

Translator
Stephen Gilbert
Gilbert Training & Translations
Beltheim-Heyweiler

ISBN 978-3-642-31732-3 e-ISBN 978-3-642-31733-0 (eBook)
DOI 10.1007/978-3-642-31733-0
Springer Heidelberg New York Dordrecht London

Library of Congress Control Number: 2012949057

Printed on acid-free paper

Springer is part of Springer Science+Business Media (www.springer.com)

MIX
Papier aus verantwortungsvollen Quellen
Paper from responsible sources
FSC® C105338

If you have any concerns about our products,
you can contact us on
ProductSafety@springernature.com

In case Publisher is established outside the EU,
the EU authorized representative is:
Springer Nature Customer Service Center GmbH
Europaplatz 3, 69115 Heidelberg, Germany

Printed by Libri Plureos GmbH
in Hamburg, Germany